3

Mechanical Man

MECHANICAL MAN

John Broadus Watson and the Beginnings of Behaviorism

KERRY W. BUCKLEY

THE GUILFORD PRESS

New York   London

© 1989 The Guilford Press
A Division of Guilford Publications, Inc.
72 Spring Street, New York, NY 10012

Printed in the United States of America

Last digit is print number: 9 8 7 6 5 4 3 2 1

Library of Congress Cataloging-in-Publication Data
Buckley, Kerry W.
   Mechanical man: John Broadus Watson and the beginnings of
behaviorism / by Kerry W. Buckley.
      p. cm.
   Bibliography: p.
   Includes index.
   ISBN 0-89862-744-3
   1. Watson, John Broadus, 1878–1958.   2. Behaviorism (Psychology)—
United States—History.   3. Psychologists—United States—
Biography.   I. Title.
BF109.W39B83 1989
150.19′432′0924—dc 19
[B]   88-24081
                                                                        CIP

For my wife, Cecelia,
and for our children,
Emily,
Hannah,
and
Nathaniel

Acknowledgments

One of the pleasures of bringing a long project to completion is being able to thank those who helped along the way. I owe a special debt to Wayne Flynt and David Vess, who encouraged me in the study of history, and to Lester Stephens, Charles C. Alexander, Charles Crowe, Robert Griffith, William Johnston, and Jules Chametzky, who have been friends and mentors. Paul Boyer's support and guidance at the initial stages of this project were invaluable, as were Jack Tager, Bruce Laurie, and Howard Gadlin's critical comments and suggestions.

I am especially grateful to Lorenz Finison and Laurel Furumoto, who introduced me to the Wellesley College Colloquium on the History of Psychology, through which I received support and advice from Michael Sokal, Bill Woodward, Larry Smith, David Leary, and Matt Hale, among others. I have benefited greatly from the willingness of Franz Samelson, John O'Donnell, Philip Pauly, Ruth Leys, Ben Harris, Jill Morawski, Charles Brewer, Brian Mackenzie, Stephen Fox, and Rand Evans to share their works in progress. I also appreciate John C. Burnham's kind words of encouragement at a crucial point. Miriam Williford deserves my thanks for her moral support as does Milton Cantor for his advice and council. I want to acknowledge a special debt to Cedric Larson, whose patient labors in the history of psychology have yielded a wealth of information which he has always cheerfully shared.

Archivists at the Yale Medical Library, the Library of Congress, the Hunt Library of Carnegie-Mellon University, the Schlesinger Library, the Bertrand Russell Archives of McMaster University, the J. Walter Thompson Company Archives, Furman University, the Ferdinand Hamburger, Jr. Archives and the Alan Mason Chesney Medical Archives of the Johns Hopkins University, and the Harvard University Archives, to name but a portion, were indispensable guides to unexplored territory.

Robert Boakes and Bernard Baars provided extremely helpful critiques at a crucial juncture of revision, as did Ronald Story, who has been a sustaining source of wisdom, good humor, and friendship. I want to thank James G. Blight for his vote of confidence and for his thoughtful and meticulous editorship, and Seymour Weingarten for his patience.

Members of the Watson family have been especially kind to a total stranger intent upon asking the most intimate of questions. I appreciate very much the hospitality and reflections of James B. Watson, Mary Watson Hartley, and Mariette Hartley Boyriven, who received me into their homes, as did Carlotta T. Watson, Joan D. McClure, and Watson's longtime friend Ruth Lieb. I am also grateful to Thomas Roe, Ruth Young, and Myrtle B. McGraw who provided additional information on Watson's life and work.

Producing a book involves the skills of many people. I would like to thank Sarah Kennedy, Marian Robinson, Denise Adler, Kathryn Abbott, Jack Cavacco, and Rich Winnick for their contributions to the design and production process.

I have been sustained in my work many times by a vast reservoir of goodwill provided by family members near and far. Wanda Ashworth Buckley, Natalie Buckley Kendrick, Vesta Davis O'Donnell, and a network of aunts and uncles have heartened my efforts, as have memories of my father, Nathaniel Cogswell Buckley, who could build a house with his own hands, and of my grandmother, Mellie Crumly Ashworth, who knew how to tell a good story.

Those friends who have helped me weather foul times, deserve my thanks in fair. Travis Absher, James C. Harper, William A. Murrah, Jack Murphy, Randall Deihl, Jane Lund, and Thomas Ulrich have been good friends, as has David Frail, who over Friday night drafts at Packard's, helped me shape a collection of words and ideas into a book.

Northampton, Massachusetts

Preface

The terms with which human beings explain and order the world they experience reflect, to a great degree, the mode of social organization dominant at the time. Historians have characterized the half-century between 1870 and 1920 as an age when Americans were engaged in a "search for order," an era when corporate rather than individual interests came to dominate the system of industrial production and the demographic balance of American life shifted from the small town to the city. The chaos, dislocation, and uncertainty occasioned by this transformation demanded remedies, which various groups—farmers, workers, embattled elites, and those who aspired to take their places—advanced in their own interest. Some scholars contend that dominant social and political classes sought to impose order and maintain their authority by institutionalizing measures of social control whereby large segments of the populace could be manipulated. Others argue that a much more subtle and pervasive process—the establishment of cultural hegemony—reveals the means by which dominant groups are legitimized—their beliefs, values, and ideas validated through public discourse.

Inquiries about these issues have focused on the rise of new professions during the last quarter of the nineteenth century. Some scholars have maintained that these new professions sought not only to establish common standards of conduct and method among their members, but also to characterize their services to society as minimizing the uncertainties associated with modern life. This biography evolved from a study of the history of American psychology—one of the most visible of these new professions. In my initial investigation I explored the development of behaviorism, which by 1920 had become the "normal science" of psychology, and its relation to the transformation of psychology during that period from a branch of philosophy into a flourishing experimental science. Behaviorism, however, did not fit the contours of "scientific revolutions" suggested by historians of science such as Thomas Kuhn. Behaviorism's rise to prominence within the field of psychology is better explained in terms of what Karl Mannheim called the "sociology of knowledge," that is, "how and in what form intellectual life at a given historical moment is related to the existing social and political forces." For, as espoused by its founder, John Broadus Watson, behaviorism had more to do with defining psychology's purpose and function than with any specific methodology beyond a seemingly rigorous empiricism.

Watson's formulation of behaviorism characterized psychology as a science whose *primary* objective should be to predict and control human behavior. Watson meant this quite literally. Throughout his career Watson gave example after example of how behaviorism could be used as an instrument of social control in the most basic sense: as a management tool, as a pedagogical method, as an advertising technique—to name but a few. Moreover, Watson's behaviorism helped legitimize the notion of efficiency as a standard of human conduct. In this sense Watson's behaviorism, which characterized *Homo sapiens* as "organic machines" and envisioned a scientifically managed society, belongs on the spectrum of progressive social thought somewhere between the Scientific Management and Technocracy movements. Above all, Watson's career as a prominent scientist, successful advertising executive, and popularizer of self-help psychology links important strands in the evolving social fabric. The establishment of professions and the rise of the expert, the growth of modern advertising, and the assault on the autonomy of domestic life are all illustrated by Watson's life and work, through which is revealed aspects of an emerging culture of consumption and self-fulfillment in twentieth-century America.

This book, then, is essentially a biography, an attempt to find threads in the flux of experience, to make sense of the welter of events through which John B. Watson lived, and by that process to illuminate his times. Watson's life is a chronicle of social mobility. Like many of his contemporaries, he followed a beckoning path that led him far beyond the world of familiar kin and community. By discerning his route we may reckon the costs exacted by such a journey and discover some of the patterns of social experience shared by those who came to shape modernist America.

Contents

— xi —

Mechanical Man

And new Philosophy calls all in doubt,
The element of fire is quite put out;
The Sun is lost and th'earth, and no man's wit
Can well direct him where to look for it.
And freely men confess, that this world's spent,
When in the planets, and in the Firmament
They seek so many new; they see that this
Is crumbled out again to his Atomies.
'Tis all in pieces, all coherence gone;
All just supply, and all Relation;
Prince, Subject, Father, Sonne, are things forgot,
For every man alone thinks he hath got
To be a Phoenix, and that there can be
None of that kind, of which he is, but he.

JOHN DONNE, *The First Anniversarie*

1

Background for Behaviorism: Growing Up in the New South, 1878 to 1900

Education, what a beautiful thing thou art!
Who knoweth what thy mighty power doth impart.
. . . we build that ladder by which we rise
From the lowly earth to the vaulted skies.

THE FURMAN ECHO, 1895

On July 20, 1900, John Broadus Watson wrote a carefully penned letter to William Rainey Harper, president of the University of Chicago. Watson was then just twenty-one years old. Only a year earlier he had graduated from Furman University, a small Baptist college near his birthplace in Greenville, South Carolina, and had managed to find a position as a schoolteacher in the tiny hamlet of Batesburg one hundred dusty miles to the south. It was a brash move for an obscure young man, so far from the centers of influence and power, to write Harper, the renowned founder of the nation's leading graduate research institution. But Watson had reached a crucial point in his life. His mother, who had raised her large family despite a hard-drinking, vagabond husband, had died a few weeks earlier. Watson was restless and ambitious, but poor. He wanted desperately to avoid slipping back into the grinding poverty that had marked his boyhood. "I know now," he wrote to Harper, "that I can never amount to anything in the educational world unless I have better preparation" at "a real university."[1]

Watson hoped to persuade Harper to grant him a scholarship to study at Chicago. His letter combined just the right proportion of self-confidence and deference that late-nineteenth-century society expected of its young men of promise. It was that Victorian quality of "earnestness" that Watson possessed in abundance. His vaguely stated ambition to succeed in "the educational world" may not have been sharply focused, but he was determined, as he put it, to "amount" to something.

Watson's desire for success was certainly not unusual. The boys' magazines and dime novels of the period still held up the intrepid image that Horatio Alger had created as a model for American youth. Yet the means by which Watson chose to achieve his ambition reflected a fundamental shift that had occurred in American society. By the turn of the century, dreams of founding an industrial empire or becoming president fired the imaginations of only the most optimistic. As frontiers of entrepreneurial opportunity began to close, the "pluck and luck" that Alger had celebrated in an earlier era could no longer insure success. Instead, many young men of Watson's generation looked to education as the passport to membership in a growing and increasingly mobile middle class. In the vanguard of this class was a new breed of professional who promised to bring about orderly growth and progress amid the confusion of rapid urban and industrial expansion. Organized around specialized bodies of knowledge, skills, or expertise, these newly founded professions established their homes within the structure of the emerging modern university.[2]

Watson glimpsed opportunities among the ranks of this new professional class and looked to the city for his future. The dynamics of industrialization, geared for an ever-increasing productivity, fostered a mode of economic and social organization characterized by a division of labor in the factory and in society at large. A new class of corporate managers and social planners emerged, one that worshiped efficiency and believed in the inevitability of material progress. Like most of his contemporaries, Watson had grown up in a predominantly rural setting permeated with traditional values and beliefs. But America was rapidly becoming urbanized and that process profoundly altered the nature and context of rural life. By the end of the nineteenth century, the village culture that had dominated the lives of most Americans was fading into twilight. Watson eventually found his destiny in the hierarchy of an emerging urban culture. But his attitudes were shaped by growing up in a rural community as it began to experience the shocks of industrialization.

Watson was born near Greenville, South Carolina, in 1878, the fourth of six children.[3] He was descended from a long line of independent landholders who had settled the back country of the Carolinas shortly after the American Revolution.[4] The South Carolina Piedmont was a region of poor farmers who maintained democratic, populist sentiments—a sharp contrast to the tidewater lowlands, inhabited by the remnants of a slave-holding aristocracy.[5] Watson's grandfather was a prosperous landowner, and each of his ten children inherited a choice section of farmland. Some of his sons became small businessmen and occasionally served as local officials, but Watson's father followed a path of his own.[6]

Pickens Butler Watson ran away from home at the age of sixteen to join the Confederate Army, where his recklessness and bravado earned him more notoriety than respect. After the war, an explosive temper and a taste for whiskey and brawls did little to improve his standing in the community.[7] But his family's patience was strained to the breaking point when he married Emma Roe. Despite Pickens's less than enviable reputation, the Watsons considered his bride's social standing to be far below their own. The marriage was the final blow to their self-esteem, and the newlyweds were, in effect, banished from the family circle. Years passed and holidays came and went, but the bonds of kinship, a sustaining force in the lives of many rural families, remained severed.

By the time John Broadus Watson was born, his family had become even more isolated. Shunned by their neighbors, they were looked upon with a mixture of contempt and pity. Pickens could not stick to farming, and the wildly fluctuating economy of the 1870s dashed any hopes of success he might have had. He tried running a sawmill some distance from home, where he would work during the week and do little else but "eat, sleep and drink whiskey" on the weekends. He became a "rover" and a "wanderer," coming home for a month or so at a time. Just as suddenly he would depart, leaving his wife to manage the farm and the growing number of children. Family tradition has it that he lived much of the time with scattered groups of Cherokees who still inhabited the remote backwoods.[8]

Such a father, whose absences were long and whose unpredictable presences were often volatile, had a lasting effect on Watson. As an adult, he rarely spoke of his father, and then only with resentment. Years later, when Watson had achieved a measure of success and fame, his father traveled the long distance to New York in a vain attempt to visit his son, whom he had not seen in decades. Watson responded by sending the old man a new suit of clothes—a gesture perhaps calculated to emphasize the gulf between their social circumstances—but he absolutely refused to see his father, thereby completing the circle of ostracism begun by his grandparents two generations before.[9]

Watson's mother, Emma, was a fiercely devout Baptist, whose hardships were matched only by her determination to overcome them. The Reedy River Baptist Church became the center of her life, and she brought up her children in strict accordance with its precepts. Rural churches within the Southern Baptist Convention were largely unaffected by the trends of liberal Protestant theology, which were considered perilously close to heresy by most of the denomination's seminaries. At least once a year, "protracted meetings," or revivals, were held. These services often lasted a week and drew the faithful and the curious from miles around. Designed to inspire believers and to save the souls of the unregenerate, the fervid pitch of the revival meetings mounted toward the inevitable climax when the congregation gathered on the banks of the Reedy River to witness the baptism of converted sinners.[10]

Her faith gave Emma Watson hope, and she made it clear to her children that their own route to salvation lay on the straight and narrow path that she marked for them. Of all her children, she singled out John Broadus Watson for a special destiny. Emma had named her son after John Albert Broadus, a prominent Baptist minister and theologian who had left Greenville the year before Watson was born to become president of the Southern Baptist Theological Seminary in Louisville, Kentucky. She fervently prayed that her son would receive the call to preach the Gospel, and through his name she gave him a legacy that would be a lifelong reminder of her hopes for him.[11]

The bewildering contrast in the characters of Watson's parents set up reverberations that echoed throughout his life. As an adult, he oscillated between poles of conformity and iconoclasm. But the patterns of family life that Watson experienced in his early childhood were, perhaps, not so unusual in rural America. The ne'er-do-well father and the long-suffering mother are almost archetypal images in

American folklore; and escape from parental tyranny or social constraints into the challenge and oblivion of the wilderness is an enduring literary theme.[12] But if a Huck Finn could "light out" down the Mississippi, Watson had to find another route of escape. In his few autobiographical sketches Watson recalled that by the time he was nine, he was "handling tools, half-soling shoes and milking cows," and that at twelve he had become a "pretty fair carpenter." He was later to define happiness as the state of being completely absorbed in activity. When anxieties threatened, he always preferred action to reflection. Although he eventually learned to channel his energies into activities considered productive by a corporate world, he ultimately looked to his rural roots for that measure of pure absorption and control that he had learned as a child. At the peak of his success in the grey-flannel world of Madison Avenue, he created an idealized rural landscape for himself on his suburban Connecticut estate. There he spent much of his spare time building barns and outbuildings and managed a working farm complete with cows, pigs, and horses.[13]

But the reality of Watson's early childhood was hardly bucolic. His family struggled on the rim of poverty. The meager existence and isolation of the farm offered little hope of escaping the treadmill of the family's circumstances. But Emma Watson was determined that her children would have the hope of something better. Watson began attending a one-room district school at the age of six. Two years later, when a private academy was founded in the nearby town of Travelers Rest, his mother somehow managed to send the children there for the next four years. The Travelers Rest Academy held daily classes in a three-room frame building that had, as its most imposing feature, a stage for the students' Friday-afternoon recitations.[14]

Emma Watson came to realize that, despite her efforts, their small farming community placed restrictions on the ambitions she held for her children. In 1890, she reached what must have been for her a bold—and for Watson a momentous—decision. She resolved to sell the farm—the only security she had ever known—and moved the family into the relatively large town of Greenville, which promised wider horizons of opportunity.[15] This was a crucial event in Watson's development. The institutions he encountered in Greenville provided a glimpse of a world beyond his experience. But it was not a world without conflict, and it demanded adjustments from those who wished to succeed in it.

The shock that Watson must have experienced when he moved from an isolated farm into a town, with its unfamiliar pace and organization, was compounded by the fact that Greenville itself was caught in the turbulence of rapid growth. The last two decades of the nineteenth century encompassed sweeping economic and social changes in South Carolina, and that process profoundly affected Greenville during the period of its transition from farming village to mill town.[16] The sleepy agricultural community suddenly awoke to find itself transformed into an industrial center when it became a primary site for the establishment and expansion of cotton mills. It was not merely the available water supply or the proximity to raw materials that drew capital for mill development to this region. Greenville's most attractive resource was its abundant supply of cheap labor.[17]

Between 1870 and 1880 Greenville's population more than doubled, and before the turn of the century doubled again as poor whites flocked to the mills from impoverished mountain coves and worn-out tenant farms.[18] The mills were initially welcomed as a providential escape from the poverty that marked the region. Many of the early mill owners adopted a paternalistic attitude toward their employees, viewing themselves as spiritual descendents of the antebellum planter class. But as new mills were established and competition increased, a new group of entrepreneurs began to dominate economically, as well as politically and socially, at the expense of the old elite.[19] One contemporary shrewdly observed that the new generation of mill owners were not "burdened with a sense of *noblesse oblige*." They were not "aristocratic, but bourgeois . . . class-conscious and money wise."[20] As Greenville became a raw, booming town, its displaced elite despaired, as what remained of the patina of cultural life that had survived the Civil War was sacrificed in the "urgency" of industrial development.[21] Colonel S. S. Crittenden, a tottering pillar of the old order, complained bitterly of "modern progress" and of a "utilitarian age" that had replaced the "romance" of an older era with "the whirring sound of turbine wheels and the hoarse whistle of the . . . mills."[22]

Although the new factories replaced old forms of peonage and chattel with new ones, they also created a basis for the emergence of a middle class. Beyond the shacks that grew around the mills, rambling bannistered and turreted Victorian houses sprouted along the new electric trolley lines that ferried factory managers and overseers to and from their jobs.[23] The mills offered the indigent, the unskilled, and the desperate a faint hope of survival. But for those, like the Watsons, who came with some resources, Greenville's economic growth provided a chance for a better life—if not for themselves, then for their children. Emma Watson moved to Greenville because she wanted her children to have educational opportunities unavailable to them in their own district. The schools of Reedy River and Travelers Rest may have met an agrarian community's need for literacy, but the new, graded schools of Greenville were organized to provide the skills and shape the attitudes required for advancement in an emerging industrial society.[24]

The Watson family's move to Greenville may have been a hopeful one for Emma, but for her uprooted children, the adjustments demanded of them proved to be difficult and unsettling. John Watson later remarked that he had "few pleasant memories of these years." He was twelve years old when he entered the seventh grade. He found the regimentation and relative anonymity of his new school to be a sharp contrast to the atmosphere of a country schoolhouse. Years later he described himself in those days as a student who was "lazy, somewhat insubordinate" and who "never made above a passing grade." As a shy boy fresh from the farm, he was unused to the social nuances observed by the sons and daughters of shop owners and mill overseers. He was often the target of classroom jokes and fought his tormenters whenever his teacher's back was turned. Perhaps it was his talent for fighting (or lack of it) that led his schoolmates to call him "Swats," a nickname that he despised but that stuck with him through college. But if Watson was bullied by his classmates, he took out his rage on the one group in Greenville that could not effectively fight back. "Nigger fighting," as he called it, became one of his "favorite going-home activities." Watson's aggressive behavior must have been exceptional indeed. He was arrested once for fighting with blacks, a step not lightly taken by white authorities in the deep South during the era of the Klan. On another occasion he was arrested for discharging firearms within the city limits. Whether Watson was a merely reckless or a genuinely malevolent youth, his adolescence was clearly a troubled one.[25]

Watson grew up in an environment where violence and racism were part of the fabric of life. An antidueling law was not passed in

South Carolina until 1881, and the carrying of concealed weapons was commonplace. Lynchings reinforced the all-pervading system of white political and economic supremacy.[26] Outbreaks of racial violence continued into the 1890s. Appeals to the federal government were made by groups in the North to protect South Carolina blacks from "organized mobs." As a young man, Watson sometimes carried a pistol to protect himself from what he called "Negro uprisings." But there were more subtle community forces that were relied upon to control black aspirations—and white ones as well. One contemporary saw the "colored churches" as a "powerful factor" in this regard. Indeed, he believed "that church membership had a good and restraining influence upon all classes."[27] Perhaps Emma Watson hoped that the church would help restrain some of her son's more rebellious outbursts. For in Greenville, the influence of the church was pervasive.

Church membership, black and white, was high in Greenville, and much of the community's activities took place within the large number of Protestant congregations. Strongest among these were the Southern Baptists. The *Baptist Courier*, the denomination's statewide newspaper, was published in Greenville, and the Southern Baptist Theological Seminary had been located there, at Furman University, until 1877.[28]

When the Watsons moved to Greenville, Emma joined the newly organized Pendleton Street Baptist Church. Although the family's religious life remained centered within the Southern Baptist denomination, the church on Pendleton Street differed markedly from the country church on the banks of the Reedy River. The larger, more urbanized church was a highly structured organization that attempted to appeal to its diverse membership. Its graded Sunday Schools for children and adults and its separate auxiliary organizations for men and women contrasted sharply with the informality of its rural counterpart. Moreover, the growth of the urban church took place in an atmosphere of fierce competition. Not only did the church compete with secular organizations and other denominations for membership, but its programs were also designed to compete with other Baptist churches within the same community. The country church, with its outdoor baptisms and brush-arbor hymnfests, began a slow decline. Large churches, with their more ornate houses of worship, were able to confer more status on those members who might eschew more worldly forms of conspicuous consumption.[29]

Watson remained a member of his mother's church on Pendleton Street until late in his college years, when he transferred his membership to the larger and more prestigious First Baptist Church. His younger brother Pickens, true to his namesake, never joined the church, but his older brother Edward was "deeply religious."[30] Watson's early religious training no doubt had an impact on the life of a man who was to develop a mechanistic theory of human behavior that denied not only the concept of the soul, but also the existence of the mind. While it is true that Watson developed a lifelong antipathy toward religion in all of its forms, he was also accused of waging an evangelistic crusade on behalf of his behavioristic beliefs. Watson's experience was typical of those of his generation who became upwardly mobile professionals. Nurtured in rural, church-going America, they came to abandon its moral certainties only to embrace a faith in material progress and a belief that the salvation of mankind would be accomplished through the achievements of science and technology.[31]

Though American society was becoming increasingly secular, it drew upon the Protestant ethic to reinforce its more materialistic aspirations. The Christian notion of "vocation," for instance, was hardly a new concept. But whereas it had traditionally referred to a calling to the ministry or to church service, by the end of the nineteenth century it had come to stand for a strong if vague impulse to serve God through the choice of a worthwhile career. For the energetic young men of the community, "service" increasingly came to mean one of the traditional professions, such as law or medicine. For the truly ambitious, the ideal of scientific research or a career in the growing fields of applied technology combined service to humanity with the march of material progress.[32]

Watson's ambitions may have been kindled by his mother and stoked by the Protestant notion of vocation, but his horizons were initially defined by the resources that his community offered. Greenville's schools were an improvement over those in rural districts, but the opportunities they provided were still severely limited. There were no public high schools in South Carolina until after the turn of the century. Students wishing to continue their education had to rely on the preparatory departments of colleges or on one of the few private academies scattered across the state. So, in 1894, at the age of sixteen, Watson enrolled as a "sub-freshman" at Furman University's "fitting-school."[33]

The denominational college was a familiar part of the landscape in small-town America in the late nineteenth century. For those without the resources or preparation to attend the elite eastern schools, the denominational college offered the most accessible route toward a professional career. These institutions were to produce the vast majority of Americans who eventually became scientists. The reasons for this lie not so much in the scientific training that was offered (which was often poor or nonexistent) but in the changing nature of the institution itself, for the denominational college was undergoing a transition in the last quarter of the nineteenth century. Church-affiliated colleges began to see themselves less as moral training grounds for aspiring ministers and more as institutions where professional skills could be acquired by those anxious to serve an increasingly secular community.[34]

Furman University was founded in 1827 as a "manual labor, Classical and English school" by Baptists who also offered theological instruction to those preparing for the ministry. When the school was moved to Greenville in 1851, the curriculum was restructured to separate the theological seminary from the traditional academic program.[35] Yet the purpose of the college for many of its Baptist constituency was to provide a moral atmosphere for the inculcation of denominational values. A Furman official warned his fellow Baptists not to trust their children's education to the state university at Columbia. The state was "of this world" and was, therefore, not "safe." Another hope offered to white Baptists after the Civil War was that denominational schools would "escape the blackening of Radicalism [Reconstruction]" that befell state institutions.[36]

During the 1890s, Furman was a college in transition. The removal of the theological seminary to Louisville in 1877 was a blow to Furman. But Greenville's growing industrialization created an impetus for more materialistic allegiances. The rising business community demanded services, and a small but growing leisure class sought culture and amusement. An alliance began to develop between the college and the new entrepreneurs.[37] During the administration of Charles Manley (1881–1897), compulsory chapel was abolished and a chemical laboratory was established. The values of competition, regimentation, and conformity were institutionalized: Intercollegiate athletics were fostered and the dormitory system grew. The old homogeneity of the student body gave way to a growing status consciousness as fraternities began to mirror the social distinctions of the outside community.[38] But in 1897, at the beginning of Watson's third year at Furman, an even greater change occurred when A. P Montague became the first lay president in the history of the school.[39]

Montague's administration was active and aggressive. He was an organizer, a fundraiser, and a builder. He reflected contemporary attitudes in an era when institutions of higher education were beginning to accommodate middle-class ambitions and to respond to the needs of the business community. Montague worked to standardize the four-year degree program and organized the faculty into committees in the name of "greater efficiency." Intercollegiate debate and athletics were promoted to prepare the body and the mind for struggle in a competitive world. He saw the future of higher education in the growth of the new professionalism and campaigned for the establishment of schools of law and medicine. Although he was not without opposition from faculty and trustees, Montague, like many other college presidents of the period, attempted to transform Furman from an institution whose primary function was the transmission of a unified culture to one that trained a new professional class to serve the diverse needs of an increasingly urban–industrial community.[40]

Watson lived at home during his five years at Furman and worked as an assistant in the chemistry laboratory to pay for his college expenses. These were "bitter" years for Watson. Although he joined a fraternity, he described himself as "unsocial" and he made few friends. A handsome lad, he wore a high starched collar and parted his carefully groomed hair precisely down the middle in the style of the period. One professor remembered him as "bright," but "more interested in ideas and theories than . . . people" and as "a person who thought too highly of himself." Watson characterized himself as an indifferent student whose study habits consisted of sporadic all-night cram sessions fortified by quarts of Coca-Cola syrup (which, in the days before federal drug legislation, included cocaine as one of its ingredients).[41]

His marks in psychology were unexceptional. These courses were taught by Gordon B. Moore, who often found himself at odds with the university authorities. Moore was eventually forced to resign from Furman because the local Baptist constituency became alarmed about his unorthodox opinion of biblical miracles. Moore's background in psychology was limited. He had taken courses from John

Dewey at the University of Chicago (from which he did not receive his A. B. degree until 1899), but these were undergraduate courses that he managed to squeeze in between his teaching duties at Furman. Moore gave special emphasis to "physiological psychology" in the classes that Watson attended. But perhaps in an attempt to placate his more orthodox critics, he liberally sprinkled his lectures with the traditional subject matter of moral philosophy, including discussions of metaphysics and free will.[42]

Watson credited Moore's courses with having led him to an interest in psychology. But, as he later reflected, it was a clash of wills between himself and his professor that "apparently" influenced his decision to earn a doctorate in psychology and philosophy. As Watson recounted this curious incident, Moore warned that any student who handed in his final exam with its pages in reverse order would automatically fail the course. According to Watson, "by some strange streak of luck," he did just that. He failed the course and was required to stay on another year. Watson then made what he later called "an adolescent resolve [to] make him seek me out for research some day." Years later, as a professor at Johns Hopkins University, Watson had his revenge. To his "surprise and real sorrow," Watson recalled, he received a request from his former teacher to be accepted as a research student. Before it could be arranged, Moore's eyesight failed; within a few years, he died.[43]

Evidence suggests that at least part of his story may be apocryphal. Watson's transcripts show no record of his having failed Moore's course (which is listed as "Ethics" rather than "Civics," as in Watson's account). But the event, as Watson recounted it, is perhaps indicative of his ambivalence toward success. Watson's constant striving for achievement and approval was often sabotaged by acts of sheer obstinacy and impulsiveness more characteristic of a flight from respectability. It is curious that Watson chose this particular professor, who had his own problems with authority and success, as a model. Perhaps Moore was a symbol of exclusion from a world that Watson wanted desperately to be a part of but could not, as yet, bring himself to enter. Watson was forced to remain another year at Furman—for which he could blame injustice or "a strange streak of luck"—and graduated in 1899 with a Master of Arts degree.[44]

In Watson's view, these years in college failed to "mean anything" in terms of his education. He was later led to believe that college "only weakens the vocational slants" and results in "softness and laziness and a prolongation of infancy." He was impatient with the notion that higher education should encourage the pursuit of knowledge for its own sake. He desired, instead, "a place where daily living can be taught." Watson was expressing an attitude shared by many of his fellow students at Furman. Members of the class of 1899 saw themselves as entering a new era; the "hero of the twentieth century" would be the "practical man" who achieved success through determination and self-control. As one student essayist put it, education was "not so much the communication of knowledge"; its primary benefit was the "discipline of the intellect . . . and regulation of the heart." Even if the "practical hero" had to "unlearn in the real world" what he learned in college, education provided the training of the will. Real knowledge was the "power" of self-control. It was the "safest route" to "success and fame."[45] A class poet penned this hymn to "The Power of Education":

> *Education, what a beautiful thing thou art!*
> *Who knoweth what thy mighty power doth impart.*
> *Let each of us ever, dear young friends, bear in mind*
> *that in this power we may fame and greatness find.*

Yet he admonished:

> *Greatness is not reached at a single bound;*
> *But we build that ladder by which we rise*
> *From the lowly earth to the vaulted skies.*[46]

Thus education came to be seen as only a calculated step in a carefully scheduled process of upward mobility. The college was no longer considered to be a community established for the transmission of cultural traditions; rather it was looked upon as the means for individual advancement. Despite his criticism, Watson viewed the socialization process of college life as a valuable commodity for ambitious young men. He later reflected that "college bred men" were "uniformly more successful" and "more likable" than others. That Watson would consider his college years to be his source of breeding implies a sharp break with his "ill-bred" family. Whether or not Watson became more likable because of it, he certainly learned that he could become more socially acceptable. He valued college primarily as

a place to grow up in—a place for breaking nest habits; as a place for learning how to make one's self friendly; for acquiring a certain *savoir faire*; as a place for learning how to keep one's clothes pressed and one's person looking neat; as a place in which to learn how to be polite in a lady or a gentleman's presence—in a word a place in which to find how to use leisure time and to find culture.[47]

It was a place where a poor boy from the country could acquire the mannerisms and trappings of a growing, self-conscious middle class. But the price of admission to this competitive and mobile world was high. A successful escape from provincialism and poverty also implied a defection for those who took flight. Assimilation into the society of urban, middle-class professionals often required, as Watson indicated, new habits, customs, and manners as well as skills. The hidden costs, however, sometimes included loss of traditions, values, and beliefs held in common by those left behind.[48]

Watson graduated from Furman in 1899, ranking fourteenth in a class of twenty-two. For the first time, he left his home and family and accepted a position, one hundred miles away, as principal of Batesburg Institute, a small private "academy" near Columbia in the central area of the state. One of his students later recalled that Watson kept to himself and avoided the social life of the community. Watson's reluctance to accept his new independence may have been reinforced by the fact that as soon as he left home, his mother fell into a protracted illness from which she never recovered. His prospects did not look promising. Watson soon grew weary of the tedium of teaching rudimentary subjects and enforcing regulations that required his pupils to be "polite, obedient . . . orderly . . . neat and cleanly at all times." But a crucial point came when, after "months and months of suffering," his mother died. Although Watson left few hints as to his reaction to this crisis, his mother's death clearly dissolved the last remaining tie to South Carolina. Soon after her funeral he took the first step on a road that he hoped would take him far beyond the world in which he had grown up.[49]

Watson had come to realize that his only hope for advancement lay in further professional training. He began to apply to some of the small number of graduate schools that then existed. He wrote to psychologist James Mark Baldwin, who was then at Princeton, but was disconcerted to learn that admission required a reading knowledge of Greek and Latin—skills that he had retained just long enough to pass his exams at Furman. Encouraged by his former professor at Furman, Gordon B. Moore, Watson then wrote to John Dewey regarding a graduate scholarship to study philosophy at the newly founded University of Chicago. Although nominally a Baptist school, Chicago had gained notoriety among Southern Baptist fundamentalists as a hotbed of modernist theology. If Watson chose to attend a Baptist-affiliated school, it was not because it was looked upon with favor by the South Carolina members of the denomination. Perhaps more appealing to Watson was Chicago's reputation as a place where president William Rainey Harper promoted the creation of practical-minded specialists to fill the ranks of a growing number of new professions.[50]

Within three weeks of his mother's death, Watson wrote his bold letter to Harper. He was "very anxious to go to Chicago," he said. He described himself as "poor," but assured Harper that he would find him to be "an earnest student." His entire future, he wrote, depended upon his ability to "do advanced work in a real university." If Harper could be persuaded to grant him free tuition or to defer his payment until after graduation, Watson argued that it would be worth his while. "I have some influence," he explained. "I have one man ready to go there [to Chicago] with me and I believe I can influence three more men to go before the year is out." It is unknown whether Watson convinced his classmates, but his persuasive powers, at least with his former tutors, must have been considerable. While his record at Furman was hardly remarkable, Watson convinced its president, A. P. Montague, to recommend him to Harper as "one of our strongest men, an alumnus who reflects credit upon his *alma mater.* . . . a gentleman of marked ability, very studious, a successful teacher, and a man of high character."[51] Although Montague may have been as eager to boost Furman as to assist Watson, the effect was the same. Harper was persuaded and Watson accepted.

In the fall of 1900, as the new century began, Watson left South Carolina for Chicago. It was a remarkable journey for a young man who had begun his life on a remote farm in circumstances of near-poverty. But Watson had learned, perhaps painfully, that mobility and adaptability were essential if he was to achieve the kind of success he aspired to. He was not unprepared for the shock of urban life that awaited him in Chicago. He had survived the turbulence of a town undergoing the stresses of rapid industrialization. Swollen by poor white migrants, torn by racial conflict, and caught up in a booming but volatile economic cycle, Greenville mirrored the problems of social and economic unrest that characterized larger urban centers in America. It was in Greenville that Watson began to adopt a vertical vision of success that led him to look beyond the world he knew for the security and acceptance for which he longed.[52]

Watson went to Chicago determined to make his mark in the world. Exactly how he would do this was unclear to him at the time. But intertwined with his vague professional aspirations were longings for the status and approval that so far he had been only grudgingly granted. Watson found his identity with those of his generation who discovered that the problems created by an expanding industrial society also provided opportunities for new professions that offered solutions to these problems.[53] When Watson entered the University of Chicago, psychology was considered to be one of the newest professions with particular promise. In order to understand Watson's decision to become a psychologist, it is important to examine the development of psychology as a profession and to consider the social context out of which it arose.

2

The "New Psychology":
The Beginnings of a Profession,
1879 to 1900

This is no science, it is only the hope of a science.
WILLIAM JAMES[1]

When John B. Watson enrolled at the University of Chicago in 1900, psychology was a profession that had been formally organized for only eight years. Yet it had already established a reputation as one of the most promising of the new professions to arise during the last decade of the nineteenth century. The founding of the American Psychological Association (APA) in 1892 was part of a larger pattern of professionalization. The last quarter of the nineteenth century saw the rise of self-conscious, sharply defined professional organizations which gave focus and direction to the aspirations of their members. The growth of these new professions reflected a fundamental change in the character of American society.

In the early nineteenth century, most people had found their place and their identity within the context of a small community. The emergence of a new industrial order after the Civil War threatened this way of life and undermined the status of old elites. But industrialization brought into being a more nationally oriented and increasingly urban-centered middle class. The specialized skills required by an urban–industrial society fostered the development of new professions in which commonly held skills provided the basis for a common outlook and identity. For those, like Watson, who sought to escape the constraints of small-town life, the acquisition of professional skills promised wider opportunities. With the establishment of the modern graduate school in the 1870s, the university became the center for the fostering and development of old professions, such as law and medicine, and the home for new ones, such as psychology.[2]

Much of the motivating force behind the professional organization of American psychology was supplied by G. Stanley Hall. It was Hall who, in the summer of 1892, called a meeting at Clark University for the purpose of establishing a national association of professional psychologists. He hoped to legitimize psychology as a science and to establish criteria for membership within the profession. Psychology had long been considered a branch of philosophy. As such, it had been slow to gain recognition as a science. Hall was one of the first generation of American psychologists who had learned experimental methodology in the laboratory and who considered themselves to be primarily allied with the natural sciences. "The new psychology" was a term used by Hall and like-minded colleagues to distinguish their rigorously empirical approach to psychological investigation from that of their more speculative brethren. By establishing a national association, Hall sought to define psychology as an experimental science and to promote the useful application of psychological research and expertise to the world beyond the laboratory.[3]

Hall had long recognized the importance of establishing professional status for psychologists and securing recognition of that status within the scientific community. Like Watson, but a generation earlier, Hall was born into a rural community. He spent his boyhood on a New England farm in the years just before the Civil War. As his parents had wished, he entered Williams College to prepare for the ministry. But he found his true calling to be the study of philosophy. In 1876, he became a doctoral candidate in the philosophy department at Harvard, where William James was encouraging the development of a "physiological psychology." James foresaw not only a change in the method but also in the "*personnel* of psychological study." Urging the application of scientific method and physiological training to philosophy, he predicted that "young men who aspire to professorships and who will bear this in mind will, we are sure, before many years find a number of vacant places calling for their peculiar capacity."[4]

Hall was encouraged by James's words. Upon completion of his studies at Harvard (earning the first doctorate in psychology given in the United States), Hall sailed for Germany, where he became the first American student to study under Wilhelm Wundt at Leipzig.[5] Wundt had made his way from the study of medicine to psychology. He had joined the rising chorus against speculative philosophy in Germany, insisting that psychology become an empirical science with a methodology of its own. He founded the first psychological laboratory at Leipzig in 1879 in order to establish a systematic science of psychology. Wundt defined psychology as the science of consciousness, the object of investigation being the immediate experience sensed by the observer. Introspection was to be the experimental method used in psychology, its goal being the reduction of consciousness into its basic elements. In adopting the introspective method, Wundt had abandoned philosophical speculation and had looked to physics and physiology, where introspection had been developed as a method to investigate light, sound, and the sense organs. By using the introspective method, Wundt hoped to discover and isolate the primary elements of consciousness and thereby to begin to discern the pattern of its structure. Above all, Wundt insisted that the study of psychology be based on data obtained through controlled experiments. By defining psychology's subject matter as observed experience, he hoped to avoid or rule out metaphysical speculations on the nature of mind that had long been the staple of traditional philosophy.[6]

Hall was but the first in a small but steady stream of Americans who sailed to Germany to study the new scientific psychology. Upon their return to America, these newly minted psychologists found that they had to create a place for themselves within the emerging scientific and professional community. They encountered opposition from two groups at the extremes of the academic and scientific spectrum. Moral philosophers, who found their positions threatened and undermined by the new scientific community, saw psychology as another example of an encroaching materialism that posed a secular threat to an already embattled religious establishment. On the other hand, those in the natural sciences, whose status had been achieved only after developing a rigorous methodology that provided concrete results, were highly skeptical of psychology's pretensions. Thus psychologists had to soothe the fears of moral philosophers while creating a basis for scientific respectability.

Moral philosophers in the first half of the nineteenth century had encouraged the study of natural science as evidence of the harmony of creation. But the spread of Darwinism transformed natural science from an ally into an enemy. Not only did Darwin's theory of natural selection challenge the biblical version of divine creation; it shook the very foundation of universal moral order. If nature was "red in tooth and claw," then it was hardly the meek who would inherit the earth. Furthermore, Darwin's *The Descent of Man* threatened to erase the distinction between man and beast and to deny human beings their uniqueness in the natural order. So when psychologists claimed that they could provide an empirical methodology for a new science of mind, moral philosophers began to give their qualified support. A new psychology based upon the study of human consciousness could give scientific legitimacy to that one remaining quality that separated man from brute and could provide an empirical basis upon which to construct a defense of free will.[7]

In 1885, G. Stanley Hall described "the new psychology" for *The Andover Review*. Echoing the pact earlier natural scientists had made with religion, Hall assured his readers, "It is the function not of psychology, but of revelation only, to give absolute truth." Striking a delicate balance between the interests of the religious and scientific communities while appealing to potential students, Hall described the new psychology as "a field peculiarly full of promise, not only of new discoveries, but perhaps still more important and inevitable restatements of old truth, and also a field in which investments of time and labor are at least as sure of quick returns in the way of positions and professional excellence as any."[8]

Only a few years earlier, Hall would have been hard-pressed to make such a statement about psychology's prospects for the future. When he returned from his studies in Germany he was unable to find a position and he feared that psychology would not be able, as he put it, "to make bread." It was then that he hit upon a scheme that would not only provide him with an income but would also promote the practical applications of psychology to a large constituency in America. Hall saw education as a promising field for the application of psychology. In Germany, Hall had observed that pedagogy had begun to receive serious scientific attention in the 1870s as the Prussian educational system underwent reform. After his return to America he had come into contact with an increasingly self-conscious group of school administrators and teachers who had begun to think of themselves as professionals and who passionately espoused in issues of educational reform. Hall began to see pedagogy not only as "the chief field of practical application for psychology" but—in view of potential effects of education on American society—as the "key to moral progress" as well.[9] While reformers considered universal public education to be a means of self-fulfillment for the individual, they urged that those so-called "liberated" impulses be channeled into "constructive" purposes for efficiency in the workplace. Nineteenth-century educators were keenly interested in finding the best means of inculcating habits of "right action" in children, and they looked to science to provide the techniques.[10]

Hall's lectures on pedagogy in Boston in 1881 were widely acclaimed in educational circles. He stressed the importance of establishing a pedagogical profession that would be based upon a scientific study of "the fundamental law of mental development." Hall perceived the mutual interests shared by educators and psychologists, and he found a receptive audience for his ideas. He was appointed to the National Council on Education and continued to maintain contact with educational groups through his involvement with what became known as the child-study movement. But his main interest continued to lie in securing an academic position. Since completing his graduate work at Harvard, he had tried, without success, to persuade Daniel Coit Gilman, president of the newly founded Johns Hopkins University, to appoint him professor of psychology. Johns Hopkins was conceived as an American counterpart to the graduate research institutions that had developed in Germany. It was dedicated to the ideal of modern scientific research and, as such, was the one place where Hall could hope to find support for establishing a program in experimental psychology. Hall, at last, was invited as a part-time lecturer in psychology in 1881; but ironically, it was his success in his popular talks and articles on pedagogy that won him his first permanent appointment in 1884—and the distinction of becoming America's first full-time professor of psychology.[11]

Gilman's choice of Hall over Charles S. Pierce and George S. Morris was a result of two factors. Since Thomas Huxley had disturbed the religious sensibilities of the Baltimore community with a stridently Darwinist lecture in 1876, the university officials were especially desirous that anyone appointed in a particularly sensitive area be circumspect in matters of potential religious controversy. In his inaugural lecture Hall was careful to point out that "deeper psychologic insights . . . are to effect a complete atonement between modern culture and religious sentiments." He further stated that "this whole field of psychology is connected in the most vital way with the future of religious belief in our land." Lest anyone misconstrue his remarks, Hall concluded his address with his belief that "the new psychology, which brings simply a new method and a new standpoint to philosophy, is . . . Christian to its root and centre." Although Hall may have been useful in stilling the troubled waters of some of Johns Hopkins's constituency, the chief element in Gilman's decision to hire Hall was the popularity of Hall's pedagogical work. This was no small matter for a struggling institution concerned about the good will of the community. Hall's influence with educational groups was not lost on Gilman. Thus the first professional position for psychology in the United States rested on the recognition of its utility as an applied science. Hall had clearly learned the value of addressing the concerns of a highly mobile and visible sector of a growing middle class by offering scientific expertise in the service of mutual professional goals.[12]

While at Johns Hopkins, Hall established the first psychological laboratory in America, which attracted students who were to make lasting contributions to the profession. John Dewey, Joseph Jastrow, and E. C. Sanford studied with Hall, as did James McKeen Cattell, who went on to study with Wundt and Sir Francis Galton and became an enthusiastic advocate of mental tests and applied psychology in general.[13] But Hall's most important contribution to psychology was the flair for organization and promotion he brought to bear on the fledgling profession. Realizing that the recognition of psychology as a science must come through "the machinery of a professional status," Hall put the machine into operation with the founding of the *American Journal of Psychology* at Johns Hopkins in 1886. Hall viewed the *Journal* as a means of defining the field of psychology. He insisted that it be used exclusively for "psychological work of a scientific, as distinct from a speculative character" and excluded all but the most empirical studies from the *Journal*.[14]

Hall's attempt to set the boundaries of psychology did not go unchallenged. William James objected to Hall's aggressive empiricism and cautioned against the exclusively experimental cast of the *Journal*.

In ignoring James's advice, Hall meant to challenge not only his former mentor's intellectual position but also his position as the chief spokesman for psychology.

Hall's relationship with James was often strained by an ill-concealed rivalry. Hall keenly felt the differences in their social origins. His struggles to escape his rural background and the intellectual restraints of Protestant morality were in sharp contrast to James's comparatively comfortable and cosmopolitan upbringing. Hall's family of western Massachusetts farmers and James's circle of genteel Cambridge Brahmins represented two extremes of the New England social hierarchy. The antagonism that sometimes strained Hall's relationship with James reflected the mounting conflict between the old intellectual elite and the new generation of technical specialists.[15]

In 1888, Hall was appointed president of Clark University, which had recently been founded in Worcester, Massachusetts. As president, Hall saw to it that psychology was established on an equal footing with the other departments. Clark promised to embody the best principles of the new higher education in America with its emphasis on advanced, postgraduate studies, and psychology gained in prestige with the establishment of a laboratory there. New universities such as Clark and Johns Hopkins were centers for the development of advanced studies. Responding to competition from these innovative new schools, some of the older universities restructured their graduate programs to attract students seeking specialized training. William James introduced a psychological laboratory at Harvard, but it was a temporary arrangement used for demonstration purposes only. The laboratory's brief tenure reflected James's distrust of what he called "brass instrument psychology"; he had, he admitted, "neither flair nor patience for experimental work." Although James respected experimentation as a means of acquiring knowledge, he was wary of what he regarded as the extravagant claims and narrow approach of experimental psychologists. He considered Wilhelm Wundt to be an opportunist who was seeking to be "a sort of Napoleon of the intellectual world . . . but . . . without genius and with no central idea."[16] Nevertheless, with the success of the laboratory at Clark, James felt obligated to offer experimental training at Harvard. In 1892, James persuaded Hugo Münsterberg (later a leading proponent of industrial and applied psychology in America) to leave Freiberg and direct a new, permanent laboratory at Harvard. "We are the best university in America," he wrote to Münsterberg, "and we must lead in psychology." Hall, in turn, felt that James was trying to undercut his program at Clark. Throughout the 1890s an uneasy and strained relationship existed between the psychology departments of the two institutions. Competition of this sort often characterized the development of graduate facilities and the rise of the university in the late nineteenth century and contributed to the proliferation of scientific psychology during this period.[17]

The most important step toward the professionalization of psychology in America took place in 1892 when Hall called a meeting at Clark for the purpose of organizing the American Psychological Association. James, who was in Europe at the time, gave cautious approval, although he confessed that he was "not particularly favorable to the organization." Perhaps owing to James's convenient absence, Hall was elected the first president. But Hall was careful to avoid any open division that could be taken as a sign of weakness by those critical of the new psychology. He made sure that the invitees represented all factions of the emerging profession, ranging from ardent experimentalists to those with a primarily theoretical interest in psychology. The object of the Association, according to its first constitution, was "the advancement of psychology as a science." The organization, it was hoped, would serve to increase psychology's visibility and recognition among more established scientific disciplines. By allying themselves with the American Society of Naturalists and the American Association for the Advancement of Science, psychologists made important contacts in the academic establishment and expanded opportunities for bringing their work to the attention of a larger scientific and professional community.[18]

The new psychology advanced rapidly during the 1890s. In 1893, James McKeen Cattell and James Mark Baldwin organized a company to publish *The Psychological Review*. The journal reflected the new professional character of psychology. It not only published current research but also reported the proceedings of the American Psychological Association's annual meetings, its presidential addresses, and notes concerning academic appointments and organizational activities of psychologists. It was also during this era that the discipline began to wrestle with its fundamental assumptions as it coalesced around competing schools of thought.[19]

The "New Psychology"

In 1893, Edward Bradford Titchener established a psychological laboratory at Cornell University. Cornell was another of the new institutions of higher learning that was dedicated to the promotion of applied knowledge. But Titchener, with his Oxford manners and formal lectures, had only contempt for applied psychology. Titchener styled himself as the only true practitioner of Wilhelm Wundt's experimental psychology in America. But Titchener's structuralism, as he called it, was an attempt to codify into a system what Wundt had merely suggested as a methodology. He relied solely on the introspective method in an attempt to describe mental states in terms of what he considered to be the basic structural components of sensations, images, and feelings. He was not concerned with individual minds or the differences among them but persisted in an attempt to discover general laws that would disclose the structure of mind. Titchener made a sharp distinction between the empirical tradition in American psychology and the experimental psychology that he rigorously pursued. He was sharply critical of what some American psychologists called "the psychology of act," which emphasized the study of the behavioral manifestations of internal mental activity. His sole ambition lay in determining the content and structure of purely mental activity. For him, the emphasis on content as opposed to act signified the distinction between pure and applied psychology. Titchener's preoccupation with models of mental structure placed him outside of the mainstream of American psychology, which, influenced by Darwin and by the social demands on American science, was concerned with the individual's "mind in use."[20]

William James articulated this position with the publication of his *Principles of Psychology* in 1890, offering an alternative to what he perceived as the static German model of psychological investigation. James conceived of the mind as an adaptive organ in which consciousness was not static, but dynamic, an ever-changing stream that enabled the organism to survive in a capricious world. Hall strongly criticized James for being "an impressionist in psychology." Ironically, James did more than Hall to gain scientific respectability for psychology. He suggested abandoning attempts to classify mental states into elements by introspection and instead dealt with a unified consciousness as a functional entity. He laid the foundation for what became known as the functional school of psychology, which, through the efforts of John Dewey and James Rowland Angell, was beginning to take shape at the University of Chicago when Watson came to study there at the turn of the century.[21]

James was broadly critical of the whole drift toward experimental psychology in America. He believed that laboratory experiments often turned out results that were "disappointing and trivial." In his view, what was needed most were new ideas. James considered the state of psychology to be similar to that of physics before Galileo: without a "single elementary law." But he was highly dubious of contemporary attempts to discover such laws. He denounced the "whole 'psychophysic law' business in Germany" as "but an idol of the cave, fit only to be kept in an intellectual museum—a sheer injury, on the whole, to the philosophy of an age." Titchener's structuralism he branded as a "pure will-o-the-wisp." As far as he could see, its only useful purpose was to "keep the laboratory instruments going." The search for the elements of mental life, he thought, was a waste of time. For James, a truly scientific psychology was one that, like natural science, attempted to discover dynamic laws by determining relationships between cause and effect. In psychology, James argued, this process depended upon the assumption of a functional unity of mind and body. James believed that the observation of everyday life provided far richer insights into mental functions than did the sparse gleanings gathered from laboratory experiments. Yet he shared at least one thing in common with most of his more experimentally oriented colleagues. The rise of modern science in America had been intrinsically connected with the notion of progress as manifested in the control of nature and an increase in material comfort. James assumed that mental science, like physical science, could develop laws and techniques for prediction and control. The control of consciousness, he believed, "would be an achievement compared with which the control of the rest of physical nature would appear comparatively insignificant."[22]

But James was no shallow booster of progress. Like many Victorians who saw their world inexorably altered by the dynamo of industrial growth, James was deeply ambivalent about the changes wrought by the advent of modern life. Although he looked to "some future psychologue" to realize his hopes for the control of consciousness, he continued to have misgivings about the direction in which psychology was growing, and he moved further away from its organizational and professional mainstream. James understood what many of his more positivistically inclined colleagues did not: the ethical implications of a deterministic world view. If human beings were merely products of social circumstances, then individual moral responsibility was inevitably undermined. He was also troubled by the apparent willingness of the new class of professional scientists to evade the larger implications of their work and to accept the routinization of scientific inquiry. He lamented that for every psychologist with a new idea there were "a hundred who are willing to drudge patiently at some unimportant experiment." The new professionals did not have room for speculative thinkers of James's range and breadth. As he explained to Hugo Munsterberg: "I am satisfied with a free, wild nature; you seem to cherish and pursue an Italian garden, where all things are kept in separate compartments, and one must follow straight-ruled walks." The scientific landscape was being mapped, sectioned off, and mined by a growing number of specialists who were replacing those intellectual adventurers who had discovered the territory.[23]

But if old frontiers were closing, new opportunities loomed on the horizon for those willing to compete within the accepted guidelines. As psychology grew as a profession during the last decade of the nineteenth century, it became well established in academia. Its success was explained by one spokesman for the new psychology in the phraseology of Social Darwinism. The systematized psychology courses and their laboratories, according to E. W. Scripture of Yale, were a "logical outcome of the conditions where the psychological work is placed in competition with the other sciences as elective courses in the great colleges." Psychologists, he advised, must offer lecturers and laboratories equal in "attractiveness to those of physics and chemistry" as a matter of "self-preservation."[24] By the turn of the century, there were forty-nine psychological laboratories in use. Eight of the major universities in the United States required psychology for the Bachelor of Arts degree, and sixty-two colleges and universities offered three or more courses in the subject. By 1904, the number of doctorates awarded in psychology increased to more than one hundred, placing psychology fourth among the sciences (behind chemistry, zoology, and physics) in the number of degrees awarded since 1898.[25]

In 1898, James McKeen Cattell wrote an article for *Science* magazine in which he described the "advance of psychology" during the preceding decade. Cattell spoke for many of his colleagues when he observed that "science has its origins in the practical needs of society" and measured psychology's advance by its increasing ability to meet those needs. But Cattell was not content to wait until the world beat a path to the laboratory door. Indeed, the growing popularity of psychology was due in large measure to the enthusiasm with which psychologists like Cattell promoted its utility. As he envisioned in the *Popular Science Monthly*, psychology offered possibilities of efficiency and order on a social scale never before realized. Allied with medicine it could determine defects in sense faculties and "degenerations" that escaped the eye of the physician. The businessman would be able to calculate the number, duration, and type of holiday that would be required in order to keep workers at peak efficiency. Mental tests could be devised that would determine an individual's suitability for a given job. But this was just the beginning. While these measures could affect an individual's behavior, they could not alter human characteristics. Cattell grandiosely envisioned combining the concepts of eugenics, which he had learned from Sir Francis Galton, with psychological techniques to devise programs that could "influence . . . the future of the race." By determining what traits were "valuable," and by the state's encouragement of suitable marriages, he argued, "degenerative tendencies" could be eradicated and the race improved much more rapidly and efficiently than by natural selection. Ultimately, Cattell believed, it was within the province of psychology to "be able to determine what distribution of labor, wealth and power is best."[26]

Cattell's vision of the application of psychology to social and economic problems reflects the transformation of the role of science in modern society. The scientific/technological revolution that began in the late nineteenth century, and is still evolving, is characterized by the incorporation of scientific research into the process of industrial production. The interdependence of universities, corporations, and governmental agencies that began to surface during World War I was clearly envisioned by psychologists like Cattell, who devoted their careers to bringing it about. But Cattell reflected the views of many contemporary reformers who put their faith in the benevolence of science rather than in what they saw as the corruption and self-interest of politics. Cattell's notion of a society perfectible by science was one that was shared by many of the founders of the profession. In their presidential addresses before the American Psychological Association during the 1890s, George Trumbull Ladd of Yale and John Dewey joined Cattell in elaborating upon the potential of applied psychology. Echoing Cattell's call for the application of psychology to "the whole conduct of life," Ladd looked forward to the contributions psychology would make "toward the practical welfare of mankind." In fields such as anthropology, criminology, sociology, and jurisprudence, Ladd envisioned, psychology would contribute to the "improved conduct and character of men." Going even further, John Dewey proclaimed psychology to be the scientific arm of democratic reform. He saw psychology as "the only alternative to an arbitrary and class view of society." Dewey was squarely within the mainstream of the American reformist tradition, and he became one of the architects of twentieth-century liberal ideology. He was fearful of the violent clashes between labor and management that had erupted during the 1880s and 1890s, but he rejected the notion that social inequality was the inevitable result of a capitalist economy. Dewey disagreed with those who would, as he put it, place the "whole industrial system under a ban." He believed that the solution to social problems lay in the "development of science and . . . its application of life."[27]

By reducing social conflict to questions of efficiency, Dewey and his colleagues reflected the concerns of an anxious middle class who could thereby turn with relief to the official faith in material progress. But the slogans of progress and efficiency often masked other motives that reflected the self-interest of the new professions. By playing on public fears of social disorder and disease and by ridiculing traditions of self-reliance as unscientific, they, if not created, at least intensified a demand for their own services. One of the key areas in which Dewey and many of his colleagues hoped to apply psychological techniques lay in the developing field of education. Since the days when G. Stanley Hall had begun to promote the science of pedagogy in America, education continued to be a subject of great interest to psychologists. As immigrants began to account for more and more of the urban population and as the institutions that reflected the homogeneity of American small-town culture began to wane in influence, the school began to be seen as the primary instrument of socialization. But the development of modern school systems and their supporting educational bureaucracies also reflected the increasing dependence of the family on professional services over which it had little control. James McKeen Cattell, for instance, regarded education as the great experiment in changing human nature "to fit the individual to his environment."[28] During the 1890s, psychologists and educators alike focused a great deal of attention on the role of psychology in the classroom. Most psychologists were in agreement about the importance of that role, but there was a wide divergence of opinion as to the degree to which psychologists should become involved in the educational process itself.

G. Stanley Hall had made extravagant claims for the use of experimental psychology in the classroom. As one of the founders and key figures of the child-study movement, he backed up his argument with data gathered from teachers' observations.[29] Other psychologists saw a vital if less direct role for psychology in education, and they cautioned teachers to temper their expectations for scientific panaceas. William James was hopeful about the eventual success of developing techniques that could be applicable to education. But he shrewdly noted that much of the interest in scientific methods came from teachers with "their aspiration toward the professional spirit." James felt that the "vague talk" about psychology was apt to be "more mystifying than enlightening" to teachers. But in his "Talks to Teachers" he was careful to point out that the social function of education went far beyond the imparting of knowledge and skills. He urged that teachers consider their primary professional task to be directed toward the "future conduct" of their pupils. The ultimate goal of education, he emphasized, was "training the pupil to behavior".[30]

Like James, Josiah Royce thought that teachers' hopes for the new psychology were "over sanguine," and he chided psychologists for "magnifying their own office." But he proposed that a position of "consulting psychologist" be created in the educational establishment as a "new division of labor" to be filled by those experimental psychologists interested in turning their expertise to "decidedly practical" matters. Royce envisioned the consulting psychologist as middleman between the classroom and the laboratory who would make experimental findings available to teachers and collect data from schools for use by psychologists. In addition, Royce argued that a school psychologist would be useful in providing ways for dealing with "defects and disorders" in the schools. As James had also emphasized, Royce believed that schools should guide and shape behavior and conduct. His litany of behavior problems may certainly be value laden, but they reflect Victorian concerns with order. Problems such as lying, "naughtiness . . . stupidity . . . obstinacy, eccentricity or precocity," Royce predicted, would yield to the expertise of the consulting psychologist.[31]

John Dewey also believed that psychologists in education should assume a middle position between the "theorist" and the "practical worker." For Dewey, the essential task was to create what he called an "organic connection" between the two extremes. Teachers should not become psychological specialists, he argued, for in doing so they assumed an analytical attitude toward students and ran the risk of treating them as mere objects. It was not enough for teachers to inculcate technique and knowledge; Dewey believed, with James and Royce, that the primary goal of education was to introduce value into the lives and habits of students. For Dewey, the school replaced the home as the principal institution for the introduction of value into society. He looked forward to a time when schools would be "managed on a psychological basis as great factories are run on the basis of chemical and physical science." The school, then, became a focal point for psychologists, since it stood midway between what Dewey characterized as the "extreme simplifications" of the experimental laboratory and the "confused complexities of ordinary life." Psychologists clearly had great hopes about the application of their expertise to problems associated with the growing demands of an increasingly complex society. Like others among the new professions, psychologists envisioned a managerial role for themselves within the hierarchy of the emerging technocratic order. But there were many problems to be solved before their hopes could be realized. Despite the rhetoric of the new psychology, the most that psychologists could deliver at the moment were promises.[32]

Although the growth of psychology as a profession was remarkable during the 1890s, at the turn of the century obstacles remained that kept the new discipline outside the mainstream of American science. There was nothing that one could point to as *the* science of psychology.[33] There was still no clear distinction within many academic institutions between philosophy and psychology; indeed, many members of the American Psychological Association considered themselves to be philosophers. Psychologists disagreed among themselves about what constituted a science. Experimentalists held that an introspective analysis of the elements of consciousness was the only path to knowledge of mind, while others felt that a search for general dynamic laws was needed. Furthermore, psychologists who attempted to establish connections between the realm of the mind and the world of experience were caught in a dualism that kept them beyond the pale of the emerging professional scientific community. There, such metaphysical concepts as "mind" and "consciousness" were considered outside the bounds of scientific inquiry. Most psychologists, however, were eager to demonstrate the usefulness of their discipline but could not reach a consensus upon a common methodology that would translate their hopes into reality.[34]

The ambivalence of psychologists in promoting the application of their science reflects a larger process of adaptation to the emerging industrial social order in America. The mingling of rosy confidence and doubt is characteristic of the Victorian bourgeoisie's response to a modernity that both threatened old verities and offered an opportunity for expanded influence. On the one hand, the broadly positivistic cast of psychology's content was a reflection of traditional American confidence in material progress. This belief was characterized by the notion that universal order was determined by laws that were knowable through scientific discovery. On the other hand, faith in scientific advancement reinforced aspects of Protestant orthodoxy that emphasized self-sufficiency and individual accountability. The drive for control of nature became fused with the notion of improvement in human welfare.

By establishing a professional organization, psychologists intended to define the parameters of their discipline and to create a role for themselves in what they perceived to be a new social order. At the turn of the century, the University of Chicago had become the focal point in the scientific and professional development of psychology. Under William Rainey Harper's leadership, the University of Chicago had been dedicated to the pursuit of scientific and social progress. It was there that James Rowland Angell had developed a "functional" psychology that emphasized the study of the mind in use and where John Dewey was exploring the social instrumentalization of scientific knowledge. When John B. Watson arrived at Chicago, he found himself in a vortex of currents that would shape the direction of modern psychology.

3

Mr. Rockefeller's University:
1900 to 1908

. . . for with money and with men the highest
ideals may be realized.

WILLIAM RAINEY HARPER[1]

When John B. Watson came to Chicago in 1900, he found a city that embodied the cultural and economic transformations that were shaping the direction of American life. Like the heroine of Theodore Dreiser's *Sister Carrie*, many in that era were attracted to Chicago as if by "a giant magnet, drawing to itself, from all quarters, the hopeful and the hopeless." It was a city where the nineteenth and the twentieth centuries collided. If the city that Watson encountered was the chaotic, ruthless, misery-ridden setting of Upton Sinclair's *The Jungle*, it was also the home of Jane Addams's Hull House, with its army of settlement workers. If it was the site of the failure of the Pullman strike it was also the birthplace of the International Workers of the World. Within the haphazard sprawl of an expanding city, Louis Sullivan and Frank Lloyd Wright were forging the ordered shape of modern urban architecture.[2]

"Chicago was the first expression of American thought as a unity," Henry Adams wrote, "one must start there." Seven years before Watson arrived, Adams had glimpsed the future at the Chicago World's Columbian Exposition. Beyond the imperial Roman facades of the exposition's White City, Adams saw the implication of planning—out of the chaos and disorder of a disintegrating society, a new order was taking shape. It was an order characterized not only by a unity of plan but also a unity of execution. It represented a collaboration of science, the arts, and industry in a rational allocation of resources to benefit the common good.[3]

The future no longer belonged to rugged individualists, but to professional specialists. Chicago's old empire builders—Pullman, Swift, Armour, and Field—were passing from the scene. They were succeeded by a hierarchy of managers who lived quietly in the privacy of their South Shore suburbs. The path to power lay not so much in the accumulation of capital as in the acquisition of expertise. Increasingly, the university became the institution looked to for the provision of that commodity.

Those visitors to the Chicago World's Columbian Exposition who dared to ride the newly invented Ferris wheel were rewarded with a panoramic view of south Chicago. The more observant passengers were able to glimpse the site of the new University of Chicago campus, whose pseudo-Gothic spires soon would rise over the grounds temporarily occupied by the fair. Opened the preceding year, the University of Chicago was the brainchild of William Rainey Harper and was brought into being with the financial backing of John D. Rockefeller.[4] Harper saw changes in the nature and focus of industrial growth and wanted the very organization of his new university to reflect that development.[5] Harper wrote to Rockefeller that he hoped the university would represent the apex in an "educational trust." He envisioned a system of feeder schools that, like their industrial counterparts, would be organized vertically to form the foundation of an educational hierarchy topped by the university.[6] The University of Chicago appeared on the educational scene full-grown at birth, designed to compete with older eastern universities like Harvard and Yale and newer institutions like Johns Hopkins and Clark.[7]

Harper's organizing skills were compared by his contemporaries (not always favorably) with those of successful entrepreneurs. He was not above using his persuasive powers and the lure of high salaries to attract faculty, often at the expense of rival institutions.[8] Harper stressed his intention to build a "useful" university. He hoped to win support by directing academic programs toward practical ends. His innovations included an extension division to broaden the university's constituency, a university press to publicize its achievements, and a system of affiliated colleges that would prepare students for advanced study at Chicago.[9]

Harper oriented the university toward the business community. He solicited the support of entrepreneurs and appointed them to the board of trustees in increasing numbers, replacing the clergymen who had traditionally filled those posts.[10] For Harper, the university was to be the home of the new professionalism. The College of Commerce *and* Administration represented not only an attempt to institutionalize the study of business as a "science" but also a response to an increasingly complex society that put a premium on administrative and organizational skills.[11]

The student that Harper wanted to attract was one "aiming to accomplish something," who worked "toward a definite plan" and was "controlled by a strong purpose" to "improve his opportunities in life."[12] Arguing that a university education was necessary for a business career, Harper promised to produce "men whose minds have been trained." By "training," Harper did not mean merely the ability to master bureaucratic tasks. For him the "advantage of college training" meant that the graduate was not content to "remain in a lower position" but continued to press forward toward something "better and higher." The function of a university education, according to Harper, was "to develop [in the student] systematic habits; to give him control of his intellectual powers; to fit him in such a manner that he may be able to direct those powers in any special direction."[13]

Harper's concept of the university reflected the changing role of higher education in the last quarter of the nineteenth century. As American society became increasingly pluralistic, the university, like the factory, had to find a way to harness and direct the energies of disparate groups. The traditional college, with its unity of purpose, had disintegrated. Uniform procedures rather than shared values became the basis for formulating administrative guidelines. As the traditional academic consensus on the definition of "quality" broke down under the pressures of a society that demanded more utility from higher education, standards of conduct were replaced by standards of production.[14] The university was not to be a place for dilettantes. The courses of instruction were highly organized and specialized. Scholarship and study in the university were tied directly to the granting of degrees. In graduate study, research was considered the "only" standard for awarding the Ph. D. degree.[15] For the faculty, research was nothing less than a "duty." "The university will be patient, but it expects every man to produce," warned Harper.[16]

Such trends did not go unchallenged by contemporaries. One of the most forceful critics of changes taking place in higher education was Thorstein Veblen. Drawing heavily from his own experiences at the University of Chicago, Veblen mounted an attack against what he perceived to be the transformation of learning into a commodity produced by "piece-rate." Yet he did not envision a return to preindustrial values. Rather, he distinguished between business methods, which he considered unscientific and wasteful, and "true" efficiency, which he believed to exist independent of pecuniary ends. Blaming the commercialization of higher learning on university presidents like Harper, Veblen considered these "captains of erudition" to be responsible for giving education a "practical bias" and thereby distracting students from "true learning." Veblen observed that university administrators were increasingly being recruited from the ranks of business executives rather than scholars. No longer was the president the first among peers, he complained; rather, he had become the employer, at the head of an impersonal bureaucracy.[17]

Other contemporaries did not share Veblen's pessimism. For them, the modern university represented society's best hopes. Harper certainly modeled his university on patterns of successful business enterprises, but the values of efficiency and production that he enthroned were part of a growing consensus among educators and reformers. Society, they believed, could shape its own destiny through the orderly application of knowledge. Harper conceived of the university as an active agent in influencing social and economic conditions. Higher learning was not to serve dispassionate and disinterested ends. As Harper envisioned it, "the true university, the university of the future," would be dedicated to "service for mankind wherever mankind is, whether within scholastic walls or without those walls and in the world at large."[18]

At Chicago this attitude was reflected in the orientation of academic disciplines. The establishment of a sociology department, for example, represented the first step in the professionalization of that field. Established by Albion W. Small, with Harper's encouragement, the department reflected the urban university's concern with immediate social issues. Small believed that the scientific study of society would reveal laws that could be used for the improvement of civilization.[19] Harper's university, however, placed strict limitations on the appropriate uses of knowledge. The new university might have encouraged social reform, but it was clearly hostile to radical solutions. Objectivity meant the avoidance of controversy. To cross the line of acceptable opinion meant to run the risk of being labeled "unprofessional"—which was itself grounds for censure and, in some cases, dismissal.[20] Emil Hirsch, writing in Small's *American Journal of Sociology*, saw the university as standing for "progress and meliorism." Although it was the duty of the academician to accept the "Herculean task of cleansing the Augean stables of civic corruption," Hirsch made it clear that a distinction must be made between fact and what he called "the half-truths urged by both conservatives and radicals." Instead of slogans, the university would offer science. Experts would replace partisans "in the upward course of humanity," and the university would be the instrument for society's improvement. It was this last point that provided the common ground for those who may have disagreed upon other issues. Ultimately uniting such antagonists as Veblen and Harper was the shared belief that science, applied to society in the name of progress, could provide efficiency and order.[21]

One of the key architects of this viewpoint was John Dewey, who came to the University of Chicago in 1894 to head the Department of Philosophy, which included the divisions of Pedagogy and Psychology. It was not by chance that these three subjects were grouped together into one department. The conceptual foundation underlying the department was Dewey's form of pragmatism, with its belief in the practical nature of all knowledge. Dewey's thought was based on a concept of action. It was an active philosophy, directed toward reshaping thought and behavior. In this sense education was philosophy at work—and psychology provided the tools with which to accomplish its chosen tasks.[22]

Education was the functional end toward which philosophy was ultimately directed; psychology was the means through which that end was to be realized. The man chosen to achieve this purpose at the University of Chicago was James Rowland Angell, who came to Harper's institution with Dewey to head the division of Psychology within the Department of Philosophy. Dewey had already begun to develop a functional concept of "mind" in describing mental functions as essentially problem-solving activities.[23] Angell developed this notion into a functional psychology that opposed the structural psychology developed by Wilhelm Wundt in Germany and became, as William James observed, the first "truly American" psychological school.[24]

For Angell, the determination of the development and operation of consciousness was at least as important as the discovery of structural elements. He considered structuralism to be unnecessarily "circumscribed and artificial," representing a complexity derived from the psychologist's analysis of experience into sensations rather than any complexity inherent in consciousness itself. Functional psychology was the study of mental operations as opposed to mental structure. Functionalism was concerned with the total relationship between the organism and its environment. Consciousness was still the subject matter of psychological investigation, but mind was not an entity distinct from the body. Consciousness was to be considered in its functional capacity to enable an organism to adjust to its environment. Sensations, Angell argued, were responses to demands made upon the organism by the environmental stimulus, with the responses varying in each specific situation. The thousands of sensory qualities that structural psychologists had struggled to identify had no actual existence apart from psychologists' descriptions—except when experience called them into being. That is to say, they exist "when there is functional demand for them." Therefore, any definition of consciousness must include some reference to the external environment of which mind is an organic part.[25]

Angell hoped that functional psychology would forge a link between philosophy and pedagogy and would provide a means of both defining and shaping consciousness. In the first psychological text to embody functionalist principles, Angell remarked that: "Inasmuch as it is mental activity, rather than mental structure, which has immediate significance for thought and conduct, it is hoped that students of philosophy, as well as students of education, may find the book especially useful."[26] Thus, when Ella Flagg Young (writing in the University of Chicago's decennial publication with Angell and Dewey) called for an applied science to be developed from experimental psychology, she was echoing members of the Department of Philosophy, Pedagogy, and Psychology in advocating a psychology for the mind in use. The application of psychology in the development of educative methods would reflect the conviction that the mind is a problem-solving organ rather than a mere storehouse for knowledge.

It would also provide a highly visible demonstration of the social applications of psychological research.[27]

The first decade of the University of Chicago's existence was a time of vigorous interaction among departments. Within the Department of Philosophy, Pedagogy, and Psychology itself there was a lively exchange of ideas and mutual contributions to work in progress. The golden era of the "Chicago School" of philosophy came between 1894 and 1904, before Dewey left for Columbia and psychologists formed their own independent department. When John B. Watson arrived on campus in the fall of 1900, the Chicago School that Dewey had conceived and organized was at high tide.[28]

When Watston later reflected upon his arrival at the University of Chicago, he recalled that he felt that he had come to the "right place" and found himself to be at home "at once."[29] But the picture that emerges is that of an ambitious, extremely status-conscious young man, anxious to make his mark upon the world but wholly unsettled as to his choice of profession and desperately insecure about his lack of means and social sophistication. He arrived on campus with fifty dollars to his name and managed to pay for his board and tuition by working as an assistant janitor at the university and as a waiter in a students' boardinghouse. Watson had initially intended to earn a doctorate in philosophy. Perhaps this was his second choice, for years later he recalled with some bitterness, his resentment at being unable to afford to enter medical school. He had no desire to practice medicine, he admitted, but he longed for the prestige of an M. D. degree to "save" him, as he put it from "the insolence of the youthful and inferior members of that profession."[30] The status that a professional degree could confer was certainly an important part of its attraction to Watson. But if he was disappointed in his desire to become a doctor, he became equally dissatisfied with the study of philosophy. Watson soon changed his major to experimental psychology. This was partly due to the influence of James Rowland Angell, who took a personal interest in Watson's career.

With experimental psychology as his major field, Watson did not quite abandon philosophy but chose it as a first minor, under Dewey, J. H. Tufts, and A. W. Moore; and he also, with Angell's encouragement, took neurology, under H. H. Donaldson, as a second minor. He also studied biology and physiology under Jacques Loeb as part of his neurology minor.[31] These were "frightfully busy years" for Watson.

His minor field in philosophy proved to be difficult for him. Two courses with Dewey failed to make an impression. Watson felt that he "never knew what he [Dewey] was talking about." A course on Kant with J. H. Tufts (which he took twice) provided "nothing" as far as Watson was concerned. It is unclear whether Watson took the course on Kant the second time because he wanted to master the material or because he failed it the first time. It is hard to imagine that he made the choice voluntarily. A philosopher who oriented his first *Critique* around the impossibility of verifiable knowledge (or of knowing that one has such knowledge) could not have been more antithetical to an aspiring experimental psychologist. Although Watson later claimed to have been influenced somewhat by the empiricism of Hume, Locke, and Hartley, his interest in philosophy was fleeting. On the whole, he later remarked, philosophy "wouldn't take hold." It is not surprising that philosophical speculation would prove to be uncongenial to a man who would come to deny the existence of consciousness. He managed to pass his exams, but, as he put it, "the spark was not there."[32]

The spark that did ignite Watson's interest was the field of animal, or comparative, psychology. He was drawn to this area of concentration partly because of his own inclinations and partly through the influence of Angell and Jacques Loeb. Laboratory experiments that involved human subjects, he confessed, made him feel "uncomfortable" and he "always . . . acted unnaturally" under those conditions, which he described as "stuffy" and "artificial." He turned with relief to the study of animals. At last he had found an endeavor in which his rural background proved to be an asset. In the animal laboratory, Watson explained, he could feel "at home."[33]

Watson initially proposed to do animal research under the direction of Jacques Loeb. Loeb had emigrated from Germany in 1891 and was invited to become a part of the original faculty of the University of Chicago a year later. He was, at first, exhilarated by the relative climate of personal and intellectual freedom that he found in America, in contrast to the restricted horizons of the German academic system. Loeb was attracted to the social philosophy of Chicago pragmatists like John Dewey but began to have misgivings about what he felt to be the reactionary tendencies of liberal reform. Dewey's philosophy, he argued, was based on an evolutionism derived not only from Darwin but also from the Social Darwinism of Herbert Spencer and from Hegelian idealism. Loeb saw this brand of liberal evolutionism as containing many of the faults of German idealism, against which he had rebelled. With its belief that "progress" was a natural process of biosocial evolution, liberal evolutionism could be used to legitimize not only social reform but also Christian metaphysics and economic *laissez-faire*.[34]

Loeb disagreed with the evolutionists' preoccupation with merely describing the process of natural selection, which they saw as a self-regulating mechanism by which organisms adapt to their environment. For Loeb, this was a passive attitude that contradicted the true role of the scientist. According to Loeb, scientific knowledge was a tool to modify and control the behavior of existing organisms and ultimately to produce new organisms artificially through biological engineering. An understanding of an organism's natural history was necessary only to the extent that it provided knowledge that would enable its behavior to be controlled.[35]

While a student of Loeb's, Watson had become convinced of the notion that man was but a biological mechanism. He listened with enthusiasm as Loeb confidently predicted that, given the correct physiological elements, an "organic machine" could soon be produced by science. In 1900, Loeb had artificially fertilized the eggs of a sea urchin. An uncompromising materialist, Loeb believed that physio-chemical explanations could account for all life processes. In popular magazines like *McClure's*, Loeb argued that instincts thought to be particularly human, such as "the instinct of workmanship" of which Veblen wrote, were different only in degree from the tropisms that cause some organisms to respond to light. Later in his life Watson acknowledged his intellectual debt to Loeb. He was certainly influenced by Loeb's thoroughgoing naturalism, and no doubt Loeb's success in writing about science in the popular press inspired Watson's later similar endeavors. But perhaps of most importance for Watson's development as psychologist was Loeb's emphasis on control of behavior as the ultimate object of scientific research. Angell, however, mistrusted the direction of Loeb's work, for Loeb's criticism of liberal evolutionism struck at the very heart of functional psychology. Eventually, Angell persuaded Watson that Loeb was "not a safe man." When Watson began research for his dissertation, it was under the direction of Angell and H. H. Donaldson.[36]

Watson's dissertation was a study of the relation between behavior in the white rat and the growth of its nervous system. There were three problems Watson hoped to solve. The first was to find out if a "systematic" account of the development of the learning process in the white rat was possible. Another problem was to discover whether a mature nervous system (which Watson defined as one in which the nerve fibers in the cortex, together with their extensions, are medullated [that is, that they possess a medullary sheath]) was necessary for the rat's ability to form and retain associations. Finally, Watson set out to find whether there was any correlation between the rat's ability to solve complex problems and the increasing complexity (measured by the number of medullated fibers in the cortex) of the rat's nervous system.[37]

Watson's study was a significant point of departure in the development of experimental psychology. The white rat, the ubiquitous laboratory animal of modern research, had been introduced to America in 1892 by the Swiss-born neuropathologist Adolf Meyer (who later became an influential colleague of Watson's at John Hopkins). Meyer persuaded H. H. Donaldson at Chicago to adopt the white rat for use in studies of the nervous system. Although Linus Kline and Willard Small had carried out the first psychological experiments with rats at Clark University, their work was primarily observational. By the turn of the century, the only psychological laboratory to employ white rats in experiments was at Chicago. Donaldson's interest had mainly been in the rat's physiology. Watson's experiments were the first systematic studies of rat behavior. But of more importance than his experimental subjects were the implications of Watson's methodology on the development of psychological research.[38]

Watson's emphasis on a systematic account of learning in animals was an attempt to differentiate the new experimental method in animal psychology from what he considered to be unscientific approaches. He criticized older investigators of animal intelligence for basing their studies on secondhand accounts of clever animal behavior. Watson praised researchers like Lloyd Morgan, who in opposition to the "anecdotal school," established the technique of observing the learning process itself and abandoned attempts to analyze learned activity.[39]

After evaluating information gathered from observing rats in mazes and problem boxes of his own design and construction, and after comparing results of dissections, Watson concluded that there was no necessary correlation between medullation and psychical maturity. The complexity of psychical life in the rat increased more rapidly than medullation.[40] Behind this conclusion lie assumptions that illustrate the influence of the functional school of psychology on Watson as well as point toward an orientation that would eventually lead him to behaviorism.

Animal psychology had always been ruled out in structuralist circles because introspection was the only method of investigation, and only human beings could attempt to observe and record their own mental activity. But for functionalists, who defined mind organically in terms of its problem-solving functions, animals could indeed provide naive subjects for observation. Watson's defined "psychical maturity" functionally, as the ability to solve problems and the ability to learn by trial and error. Thus he concluded that psychical maturity was not defined physiologically and did not depend on the physiological state of fully medullated nerve fibers; rather it was determined behaviorally by the functional criterion of mind as a problem-solving organ. This did not mean that there was no difference between problem-solving activity at the point of psychical maturity and the problem-solving activity of the adult. In his one analogy between mice and men, Watson suggested that the difference between the adult and the child at psychical maturity (the point at which it becomes teachable) is that the child lacks the learning experiences that the adult had acquired with age. Not lost on Watson, as he watched the rats negotiate his mazes and puzzles, was the extent to which these learning experiences could be controlled.[41]

Watson worked "night and day" on his experiments. It was during this time, he later recalled, that he began to establish "work habits" that stayed with him all his life. He was learning that in the new professional world, scientists were not exempt from the rules of competition. Survival and advancement in that world demanded the ability to discipline one's self to a regimen of production and efficiency. Watson was driven by ambition, and he accepted the challenge willingly, but not without cost. His obsession with achievement reflected deep anxieties about failure and success. Throughout his life he courted both with equal determination, and that conflict brought him, more than once, to the breaking point. The year before he completed his doctoral work, the machinery of Watson's carefully constructed routine shattered when he suffered what he called a "breakdown." Weeks of insomnia followed by a period of enforced rest during which he could sleep only in a well-lighted place were the manifestations of what he later described as "a typical *angst*." Returning to his work after a sudden recovery, he saw the episode, in retrospect, as one of his "best experiences." Not only did it teach him to "watch his step"; but he claimed that his acceptance of a "large part of Freud" several years later sprang from the events during this period.[42]

Watson's experience is especially revealing when placed within the context of late-Victorian American society. George Miller Beard, an American neurologist, had coined the term "neurasthenia" in the 1880s to describe various states of emotional distress characterized by a morbid introspection that resulted in a paralysis of the will. By the early twentieth century, physicians and culture critics had proclaimed neurasthenia to be the disease of the age, affecting professional men and their families in increasing numbers. Whether or not nervous disorders were actually on the rise, contemporaries believed it to be the case. Most observers blamed the proliferation of emotional illness on the frenzied pace of modern life. It was the struggle for existence that overtaxed the mental powers of the professional classes, among which the ability to compete depended upon intellectual rather than physical strength. At the turn of the century, the prevailing assumption among American therapists was that emotional strength needed to be husbanded. The recommended cure for nervous prostration was protracted rest and isolation in order to rebuild one's dissipated emotional energies to the point where one could exercise the measures of self-control needed to keep them from being depleted. The diagnosis and treatment of nervous disorders were based on the notion of psychic scarcity: One had limited emotional reserves that could easily be overdrawn. The connection between repression and neurosis was only dimly perceived. But as the expanding economy began to be based more and more on a culture of consumption, a new faith in psychic abundance began to take shape. The mind, particularly the unconscious mind, began to be seen as a source of tremendous energy whose potential could be released. If Watson believed his *angst* to be "typical," he was echoing the anxieties of many of his contemporaries in their confrontation with the demands of urban–industrial life. That he valued the experience for teaching him to "watch his step" reflects the late-Victorian preoccupation with the exercise of self-control over scarce emotional resources. Seen in this light, what Watson meant by the acceptance of a large part of Freud becomes more understandable. For Watson hardly abandoned the notion of psychic scarcity but went on to develop a theory of the emotions whose development depended entirely on external conditioning. What behaviorism and psychoanalysis ultimately had in common was a belief in the plasticity of human nature and in the temporality of social institutions. This signified a break with the values of Victorian society. Standards of conduct that had been set and enforced by community and church were being replaced by an ethos that stressed self-fulfillment and personal gratification. Far from being liberating, however, this shift had the effect of casting the individual adrift in a sea of shifting values that were determined more and more by styles of consumption.[43]

Watson's solution to his own emotional crisis was a consuming careerism to which all other considerations became secondary. Watson was later fond of quoting James's admonition to stand by one's truest self and to forsake all other possibilities. Ironically, James characteristically refused to define the self narrowly and represented in his own time a vanishing type of intellectual, shaped by both the pre-Victorian notion of vocation and the twentieth-century demand for specialization. Watson, on the other hand, viewed the self as defined by the choice of one's career. He conceived of life choices not as a series of options, each with their own possibilities, but as occupational roles acted according to a script that one carefully selected early in life. Any ambivalence about the role one chose in life, he believed, was to be repressed once a selection was made. All other considerations were to be "thwarted," as he put it, even though the desire for rejected choices still persisted.[44]

Watson's sense of assuming and performing roles reflected the concerns of a society in which social mobility was becoming part of the folklore of success. The climb up the social and professional ladder required keen abilities of observation and adaptation. Watson depended upon professional achievement for his self-esteem, and he was inclined to take criticism and competition from his colleagues as personal affronts. When Watson received his Ph. D. in psychology in 1903, he was the youngest to have earned that degree from the University of Chicago. Yet he felt what he called his "first deep-seated inferiority," which grew into years of jealousy, when he was told by Dewey and Angell that his doctoral exam was inferior to that of Helen Thompson Woolley, who had graduated two years before. Watson was often a willing mentor to his female students, but he was extremely uncomfortable with women as professional peers.[45]

Watson, however, had little reason to feel threatened. His attitude was certainly not in the minority among psychologists. It was his work, rather than Helen Thompson Woolley's, that received the imprimatur of the department. Dewey brought Watson's work with the white rat to the attention of James McKeen Cattell, then the editor of *Science*, and Watson represented the department by presenting a summary of his research at a regional meeting of the American Psychological Association.[46] Upon Watson's graduation, H. H. Donaldson encouraged him to apply for an assistantship from the Carnegie Institution. Although his application was unsuccessful, Watson was offered other alternatives. He could have gone to the University of Cincinnati as an instructor, but Dewey and Angell wanted him to stay on at Chicago. Angell's failing health made it imperative that he have an assistant to take over some of his research and teaching, and Watson was considered by both men the most "thoroughly competent" for the position. In the fall of 1903, after much wrangling with the parsimonious administration, they were able to offer Watson an instructorship, which he accepted at one thousand dollars per year.[47]

During Watson's first year as an instructor at Chicago, the psychologists in his department began to assert their desire for a separate identity within the university. Angell had declared psychology to be "largely independent of philosophy" and considered it to be closer to biology than to any other academic discipline.[48] When Dewey resigned in 1904 to go to Columbia, Angell saw his chance to counter the "depressing influence" of Dewey's departure and at the same time create a measure of independence for psychology. Angell wrote to university president Harper proposing the establishment of a separate department of psychology, reassuring him that the expense of such an undertaking would be "entirely practicable." Harper was convinced, and a psychology department was organized, with Angell as administrator and director of research in general psychology and with Watson conducting laboratory work in animal psychology and teaching the introductory course in experimental psychology. Angell also drew on faculty outside the department to give instruction in related areas. From the philosophy department George Herbert Mead offered courses in comparative psychology, A. W. Moore taught educational psychology, and E. S. Ames gave instruction in the psychology of religion; William I. Thomas from sociology offered work on the psychology of primitive peoples.[49]

Thus Angell asserted psychology's independence by severing its organizational ties with philosophy. Angell slipped comfortably into the role of administrator. Assuming more and more administrative duties at the University of Chicago, he later became chairman of the National Research Council, then director of the Carnegie Corporation. He eventually became president of Yale University.[50]

As department chairman, Angell was constantly forced to petition the university administration for funds to supplement his modest research budget. Harper had to be "sold" on the practicality of such expenditures, and Angell recognized this need when he backed a request of Watson's for laboratory funding. Not only should Watson's work be supported for its "unquestioned intrinsic value," Angell argued, "but also because it is enjoying a great 'boom' just at present among the universities, and to be found deficient in it is to reflect upon our organization and equipment." Harper was assured that he could "feel safe" in backing Watson. Watson had already secured an "enviable repute" among his colleagues with a "flattering" review of his thesis in *The Nation* and "commendatory notices" in the journals. Angell did not fail to point out that Watson was a valuable commodity to the university. He reminded Harper that Watson had already been wooed by several institutions "not forgetting . . . Hopkins."[51]

Harper was not one to allocate funds rashly, and Watson only managed to get a promise of funding for the following year. Watson felt "very much hampered" in his research, he complained to Harvard's Robert M. Yerkes. He had no facilities to keep his animals and no funds to maintain his "menagerie" even if space were available. He felt "loaded down" with courses and with the responsibility of keeping track of all the laboratory business records.[52] But Watson earned Angell's admiration and was appointed his assistant. Angell considered Watson "extremely skillful" in his field and hoped that he would be able to initiate work in comparative psychology at Chicago "under laboratory conditions." Watson was "a man of firm personality and high character," Angell wrote to Harper, adding that he was sure Watson would "do us credit"—a statement not lost on a man skilled in evaluating balance sheets and shrewdly aware of the public-relations dimension of his enterprise.[53]

Watson's career was greatly enhanced by Angell's confidence. In September 1904, he was chosen to represent his department by presenting an address at the Congress of Arts and Sciences at the Louisiana Purchase Exposition in St. Louis. But he was quick to demonstrate an independence of mind. Following the line of an editorial he had recently published in the *Journal of Comparative Neurology and Psychology*, Watson attacked the tendencies toward what he considered to be unscientific generalization in current research.[54] He felt "vague and dissatisfied" when he read general studies of the "'mental processes of animals.'" He proposed more specialization and "more exact and more restricted studies." For Watson, nothing short of a "division of labor" would elicit the results he desired. Areas of specialized study should be restricted to experts. The comparative psychologist without neurological training could "find enough to do on the strictly psychological side." In developing his own approach to comparative psychology, Watson felt that, although his tests were too few and inadequately controlled, the conclusions drawn by the English psychologist L. T. Hobbhouse were "in line with progress." Summarizing these conclusions, Watson emphasized that animals seem to learn by "attention to a simple sequence of events" that either "gratifies or hurts them"; that behavior can be described as an action directed to an external change in the perceived object rather than a motor reaction to a perception; and that there seems to be "no natural tendency to learn by perception." Watson called for experimental work that would reconstruct existing concepts of memory, imitation, and reasoning. This work, he emphasized, should "always be in closest touch" with studies of human infants. He pointed out that "almost nothing" was known about the "uses" of senses in animals and advocated the testing of animals whose senses had been systematically eliminated. Recounting experiments that revealed the intimate relationship between physiological structure and behavior, Watson maintained that "the study of structure and function must go hand in hand" in order to develop a complete knowledge of behavior in higher organisms. This was a direct challenge to structural psychologists, who relied on introspection to arrive at an analysis of perception. In effect, Watson excluded them from contributing to an understanding of behavior, emphasizing, as he did, the necessity of specialized scientific skills, especially neurological training, in order to conduct investigations.[55] Criticizing earlier investigations as imprecise and unscientific, Watson represented a new generation of psychologists who called for even more specialization and allied themselves with the biological sciences rather than with philosophy. His emphasis on behavior rather than perception reflected the influence of Chicago functionalism and pointed the way toward behaviorism, which would at once define more narrowly the subject matter of psychology and expand the possibilities of its application.

Watson's professional concerns with order belied a tempestuous and sometimes chaotic personal life. His scientific work exhibited a cool detachment, but he had difficulty placing limits on what seemed to be an impetuous and romantic temperament. Yet if Watson courted chaos in his personal relationships, it served to provide him with the emotional distance that he required. For below the surface of his romantic posturing, which he played out until its inevitable collapse, was a deep cynicism and a profound distrust of emotional intimacy. The relationships that he chose involved women who were young, impressionable, and, initially at least, awed by him. Soon after he received his degree, Watson fell in love with Vida Sutton, a student at the University of Chicago. She rejected his affections, and Watson soon turned his attention to Mary Ickes, a nineteen-year-old student at the university.[56] Her brother, Harold Ickes, a lawyer who went on to become Secretary of the Interior in Franklin Roosevelt's New Deal cabinet, was then active in Chicago reform politics. Born in a small town in western Pennsylvania, Mary, like Watson, had been raised in a devoutly Protestant household. Mary's father, who was rarely at home, sent the children away to live with relatives in Chicago upon their mother's death.[57]

As family legend has it, Mary was a student in Watson's introductory psychology class. She developed a crush on her professor and during one long exam wrote a love poem in her copybook instead of answers to the test questions. When Watson insisted on taking the paper at the end of the quiz, Mary blushed, handed him the paper, and ran from the room. The literary effort must have had its desired effect. But the courtship that ensued was hardly blissful.[58] Seventeen years later, Watson revealed the story in a letter to psychiatrist Adolf Meyer. In a perhaps self-serving narrative, Watson explained that when he became involved with Mary Ickes, she was embroiled in a bitter struggle with her brother, Harold. Mary's mounting debts had moved her brother to write her "furious letters," and Watson, not wasting an opportunity to indulge his penchant for melodrama, added that he actually feared for her physical safety. His sense of chivalry thus aroused, Watson stepped in to rescue his lady in distress. "I was young and foolish and Southern," he wrote, and he and Mary were secretly wed on December 26, 1903. This was "long before," according to Watson, "any sexual intercourse was even attempted." But the fact of their marriage was kept secret, and when Harold Ickes decided to remove his sister from college and send her to an aunt in the East, Mary was forced to comply with her brother's demands. While she was away, Vida Sutton returned to Chicago and told Watson that she had made a mistake and was in love with him. Watson told her of his marriage, but they continued to see a "good deal" of each other, even though their relationship, Watson emphasized, was "exemplary" from "a legal standpoint." After several months, they finally realized the hopelessness of their situation and "gave each other up." Watson sent for Mary, and they were publicly married in the fall of 1904. Watson confessed everything to his bride, even though he realized that the situation did not provide "a very good foundation for marriage."[59]

The marriage between Watson and Mary Ickes proved to be painful and dissatisfying to both parties and eventually ended amid sensational publicity sixteen years later. Their troubles were compounded by the bitter rivalry that developed between Watson and his brother-in-law. Watson bristled at Harold's insistence on managing Mary's financial affairs. Ickes, however, held nothing but contempt for Watson. A self-styled "curmudgeon," Ickes minced no words in expressing disapproval of his sister's marriage. He had "sized" Watson up as "a selfish, conceited cad." "You are right in thinking I do not like you," Ickes wrote to Watson a few months after the marriage, "in fact I do not even respect you. I choose my friends from men." He bitterly regretted having given his consent for Watson to marry his sister. "Inquiries" he had made concerning Watson convinced him that Watson was not liked or respected among those he considered to be "men" at the university and alleged that the women referred to him as "that Watson." Accusing Watson of being "disloyal" to his sister during their engagement, he warned Watson to "take care these silly confidantes do not publish your confidences broadcast or I may find out more truth about you." Although he did not "blame" Watson "for being poor," Ickes accused Watson of not adequately supporting his wife and of allowing his own "selfish interests to be paramount." For Ickes, Watson violated the code of gentlemanly behavior expected of "Northern men" like himself, who found it difficult "to grasp the character of a man who will fight from behind a woman's skirts or carry a pistol about with him." Besides being suspicious about Watson's southern roots, Ickes was more than contemptuous of what he considered to be Watson's social pretensions. Ickes disapproved of Watson's friends and refused to associate with them. Not without ambitions himself, and despite the fact that his own family's circumstances were modest, Ickes nevertheless told Watson that he was mistaken if he appropriated to himself "the reputation for breeding that is popularly supposed to belong to people from the South." Asserting that "breeding is individual and not sectional," Ickes accused Watson of being "singularly lacking in respect to real good taste, courtesy and breeding." Furthermore, he admonished, "many men who know enough to take off their hats to ladies are not necessarily gentlemen."[60]

Whether or not Ickes was justified in characterizing Watson as an unprincipled social climber, Watson did little to smooth Ickes's ruffled feathers. But Ickes was not the only one whom Watson provoked to the point of outrage. Watson, Ickes grumbled, was constantly complaining about "expense, cost and money." This tendency often tried the patience of Watson's superiors at the University of Chicago.[61] He was at times impetuous in his dealings with the university administration over allocations, and he often rankled bureaucratic sensibilities. When Watson submitted a request for additional laboratory equipment, some ill-considered remarks contained in the request were taken by William Rainey Harper to be either an "indication of insanity, or intentional impertinence." A personal intervention by Angell and a formal apology by Watson were necessary before Harper's injured feelings were soothed.[62] For Harper to consider a remark to be "insane" was certainly no idle rejoinder. Watson was capable of being ingratiating when it was in his interests, but he also achieved a notoriety for a remarkable lack of tact. Watson certainly craved success and worked hard to achieve it, but his ambivalence about his accomplishments often surfaced in actions that seemed to sabotage his best efforts. Yet, no one could argue with his determination.

Watson was anxious to get on with his research and found the teaching load, limited facilities, and lack of funds constricting at

Chicago. With the support of Angell, H. H. Donaldson, and James Mark Baldwin, he again applied to the Carnegie Institution for a research grant. His application reflected his current thinking about the state of psychology. Criticizing psychologists who contended that animals may be assumed to have a "mental life" on purely logical grounds, and heaping contempt on those whose "naive assumptions both psychological and metaphysical" were untested by careful experiment, he proposed to carry out experiments designed to show the *functions* of the various sense organs of animals in the learning process.[63]

Although he was once again turned down by the Carnegie Institution, Watson applied for a leave of absence, hoping to spend the spring and summer in research. The move, for Watson, was a risky one. The leave of absence was without pay, and, incurred the wrath of his brother-in-law for pursuing "selfish interests," Watson took his pregnant wife to Baltimore, where he studied the aseptic method in animal surgery at Johns Hopkins University under William Howell.[64] He had hoped to find employment in Baltimore to support his studies but soon found himself broke and in debt. In desperation, he wrote Angell, who, in turn, urged Harper to authorize a loan for Watson from university funds. Describing Watson as a man with "unmistakable promise," he assured Harper that he would find the money "well invested." "After all," he argued, "fine human beings are the best paying investment and I think Watson belongs in this class." Harper evidently agreed with Angell and forwarded Watson a check for two hundred dollars as a loan for one year at four percent.[65]

The assistance came none too soon, for Watson's wife had given birth to a daughter only a few days earlier. Watson's reaction to assuming fatherhood was ambivalent at best. Advising a colleague about marriage, he cautioned that "it will interfere with your work a little at first," but he felt that it was the "right thing" to do because "it will pay in the long run." But for a time, at least, the new member of Watson's family held her own against the laboratory animals. Watson admitted that he found a baby to be "more fun to the square inch than all the frogs and rats in creation."[66]

But upon returning to Chicago, Watson threw himself into his work. In addition to his regular teaching and laboratory duties, he carried out experiments on imitation in monkeys, edited (with help from his wife) an edition of the *Psychological Bulletin*, and organized a meeting of the western branch of the American Psychological Association.[67] Not only was Watson's reputation for experimental work growing, but his skill as an organizer and administrator was being recognized and sought after within the profession.

It was during this period that Watson was engaged in research that would bring him national notoriety for the first—but hardly the last—time. In a long and painstaking experiment, Watson hoped to prove that "kinaesthetic-cutineous [*sic*] sensations" were responsible for the associative, or learning, process of the white rat. His method was to compare the responses of normal rats with those of rats who had been deprived successively of the senses of vision, smell, and touch.[68] When the results of these experiments were presented to the American Psychological Association later in 1906, Watson felt that he had stirred up a "nest of hornets." *The Nation* editorialized that such "cruelty was so nearly purposeless as to be wholly unjustifiable." While admitting the usefulness of vivisection for medical "progress" and the "prolongation of human life," the editor felt that a protest must be made against the "torture of animals for merely trivial investigation." Watson was depicted in newspaper stories as a cynical, aloof, detached "rat scientist" with sinister and thinly veiled sadistic motives. Ambivalence about technological progress was often reflected in the image of the scientist in the popular press. A Pasteur could be hailed as a messiah, but the ghost of a Dr. Frankenstein was never far away in the shadows.[69]

Watson's colleagues rallied to his defense and to the defense of the profession. Angell urged James McKeen Cattell to print a rebuttal in *Science*, and James Mark Baldwin justified the importance of the experiments in a reply to *The Nation*.[70] In the published results of his investigations, Watson emphasized the overall direction of his work and its implications. His study, he claimed, was aimed not merely at examining the particular species of animal studied, but at understanding mind in use and in establishing a functional comparison of the sense-organ processes in animals with those of human beings. Comparative psychology had drawn criticism from structural psychologists, who held that one could never investigate the mental life of animals because one was barred by the introspective method from studying its content. For Watson this criticism had "no significance," because his view of "mental processes" was from a purely functional standpoint. His interest lay in how an organism reacted to its environment. This could be determined not by introspection, but by observation of behavior. By observing animals whose senses had been systematically removed, one could determine, by their behavior, the function of the senses. Thus Watson moved away from the traditional subject matter of psychology—the study of the mind—and took a large step toward establishing a new science of behavior.[71]

Watson did not confine his views to his colleagues but very early in his career discovered a receptive audience in popular magazines. In a 1907 article for *The World Today*, he argued that mind was just as much a result of the process of evolution as the body's structure. Since the mind was an adaptive organ, an understanding of how it worked involved the study of its adaptive function, not its structural makeup. Like the companion of Sir Arthur Conan Doyle's Dr. Watson, he maintained that the only way we have of knowing what a man is thinking is "by noting carefully what he does!" Thus, he argued, one could use the same method to study the minds of animals or human beings. Watson minimized the importance of language as a factor distinguishing human beings from animals. Language, he believed, was merely a more elaborate and complex category of behavior. Watson had great hopes for the implications of his work. He predicted that the study of behavior would produce "fruitful results for the guidance of human conduct."[72]

Such long-range goals, Watson argued, required laboratory facilities that would provide continuity for observations and experiments. He had editorialized for the establishment of an experimental station outside the confines of city laboratories that would allow for the continuous observation of successive generations of animal species. This facility could be used, he argued, to answer such long-standing questions as whether mental traits are inherited or whether there are sexual differences in learning ability. Data gathered by experts in a controlled system would replace earlier, unsatisfactory, and unscientific conclusions based on anecdotal evidence. But of more importance, Watson's proposal was aimed at establishing a role for psychology in providing solutions for issues of major social concern. At a time when neo-Lamarckian concepts of biological determinism were being challenged by cultural relativists in the social and behavioral sciences, psychology, he predicted, could devise new ways to account for (and to control) fundamental differences in behavior.[73]

Although the establishment of an experimental station was not immediately forthcoming, Watson was selected by the Carnegie Institution to establish a station to observe seagulls in the Dry Tortugas, a group of small sun-baked coral islands in the Florida Keys.[74] In what was to be the beginning of a series of experiments there, Watson spent the summer of 1907 under less than ideal conditions. The climate was unbearable, he complained to Robert Yerkes. By night he was "dog tired," with scarcely enough energy to write. By observing more than five thousand young birds he hoped to get a "genetic statement of the instincts *a la* Lloyd Morgan." In regard to this he found that instincts imperfectly present at birth are in fact perfected by habit, that is, through "trial and error." Only after "hereditary coordinations" had been established, he felt, could the study of the psychology of any species become possible.[75]

Returning to Chicago in time for the brith of a son, Watson declared that "two kids are enough, Teddy [Roosevelt (prolific father and advocate of the "strenuous life")] to the contrary." Using "everything modern" in the way of raising babies "according to Holt," Watson was not bothered "in the least" by hearing his children cry.[76] Perhaps he considered this to be a mark of advanced parenting. But his temperament as a father was hardly warm. His daughter recalled that the only time her father was physically affectionate toward her was when he departed for Europe during World War I—and then he merely kissed her on the forehead. The family's happiest times, as she remembered them, were during her early childhood, when they would spend the summers in Canada. Despite his complaints about finances, Watson had purchased an island on Stony Lake in Ontario and over several summers had, with the help of his family, built a spacious cottage, complete with flagstone fireplace and paneled walls. Watson loved an active life, and contact with his children centered around physical activity. He taught his daughter to swim and dive, but his son, the younger child, seldom received the same attention from his father. In his family and in his social life, Watson always preferred the company of women.[77]

This predilection often put his marriage under severe strain and came close to causing a scandal at the University of Chicago. Sometime during 1906 or 1907, Vida Sutton, the object of Watson's earlier infatuation, returned to Chicago. Watson's brother-in-law, Harold Ickes, became suspicious and hired a private detective, who reported that Watson was meeting Miss Sutton regularly. The motivations that moved Ickes to take such extreme measures were not entirely ones of concern for his sister's happiness; it was precisely at this time that Ickes himself was involved in an extraordinary affair that must have incited his suspicions about Watson. While a boarder in a large Chicago house, Ickes had become involved in an affair with the wife of his landlord. He eventually persuaded her to divorce her husband and marry him. But all of this took place with an intrigue and secrecy that placed a heavy burden on his Victorian sense of morality and order. His difficulty in facing these contradictions in his own life no doubt partly explains the savagery of his attacks on Watson. Although Watson denied his brother-in-law's charges, Ickes went to Henry Pratt Judson (who had become president of the University of Chicago upon William Rainey Harper's death) and insisted that Watson be fired. According to Watson, Judson investigated and found no truth in the accusations. Even so, Ickes demanded that his sister sue for divorce. At that point, James Rowland Angell intervened to keep the marriage together. Watson later reflected that "it would have been by far the wisest thing had he let us separate."[78] That Watson credited Angell with saving his marriage is revealing, for either Angell exerted an extraordinarily powerful influence on Watson, or Watson's attitude toward his marriage was extremely passive. In either case, by blaming Angell's poor judgment because he did not "let" them separate, Watson showed an unwillingness to assume responsibility for the decision.

For Watson to have had his personal difficulties flung so dramatically before his colleagues and superiors must have created an uneasy situation within his department. In addition, he continued to feel inadequately compensated for the time and effort that he put into his work. Watson had accumulated over two thousand dollars in debt in order to get his degree. He had married with the hope that his advancement would be rapid, but he remained "tied" to an instructorship at low pay until he skillfully used outside offers to negotiate raises in his rank and salary. Even so, he complained to E. B. Titchener that he was still on financial "pins and needles." He "slaved" in the laboratory, doing everything from installing electrical wiring to janitorial work and had to do "more hack work on the outside than good work on the inside in order to make both ends meet." Although he had published a good deal, he felt that his research was done "hastily" and, as he lamented to Titchener, "at the expense of my nervous system."[79]

Accordingly, when the American Psychological Association met in Chicago early in January 1908, Watson was open to offers from other institutions. His professional reputation had grown during his years at Chicago. Angell had nothing but praise for him. H. H. Donaldson considered him to be the "least adequately recognized man" in psychology. But despite prodding by Angell and others, Henry Pratt Judson could not be moved to raise Watson's salary or rank. Angell faced the prospect of losing Watson with great reluctance. But he had a paternal interest in Watson's career and was determined that nothing should stand in his way. "I shall never let him go if I can help it," he wrote, "but I am too fond of him to let my selfish interests outweigh any opportunity to get him proper advancement, and if I cannot get it here, I shall always try for it elsewhere." So, with Angell's blessing, Watson became the object of an elaborate courtship from the University of Chicago's rivals. E. L. Thorndike approached Watson with an offer to come to Columbia (where Dewey had gone four years earlier), but the most serious contender for his services was Johns Hopkins University in Baltimore.[80]

Watson's reputation, however, was not untarnished. He had earned the enmity of some psychologists because he seemed to thrive on controversy for its own sake. Harvard's Robert M. Yerkes thought that Watson, at times, resorted to "unnecessary" criticism calculated to "provoke antagonism." Watson, he warned, had a habit of "looking for trouble." Yerkes feared the development of a cult of personality in animal psychology "if Watson's lead" was followed. Yet Yerkes's opinion was no doubt influenced, at least in part, by his failed attempt to land the position at Hopkins for himself. Herbert Spencer Jennings, a zoologist at Johns Hopkins who had felt the sting of Watson's criticism, also had reservations. He believed that Watson was inclined to take his ideas in "too absolute a way" and considered his position to be "strangely wooden and narrow." Yet he thought that Watson would be a "great acquisition" for Johns Hopkins and "would rejoice" to have him as a colleague. Despite Watson's drawbacks, he was perceived as a psychologist whose career was on the rise. An engaging simplicity endeared him to those who, like Angell, were willing to dismiss his faults as youthful exuberance. Above all, Watson had, as H. H. Donaldson observed, "that gift of Heaven of getting things done."[81]

James Mark Baldwin, head of the Department of Philosophy and Psychology at Johns Hopkins, believed that Watson would be a valuable asset to his program. President Ira Remsen agreed. "It is clear to my mind," he wrote to Baldwin, "that he is the man for us." He considered the acquisition of Watson to be an "important case" and stressed to Baldwin that, "we ought not to lose a man like that." Watson had the advantage in his negotiations with Baldwin, and he did not hesitate to make the most of it. He adroitly used Columbia's interest in him to encourage Johns Hopkins to bid high for his services. As a result, in March 1908, Watson accepted President Remsen's offer to come to Johns Hopkins the following fall as full professor at almost double the basic salary he had received at Chicago.[82]

Watson was but twenty-nine years old when he accepted the position at Johns Hopkins. Just eight years before, he had come to Chicago with little more than an ambition to "amount to something." Yet by 1908, he had established a professional reputation far beyond those of most of his peers. His enormous capacity for productive work and a pugnacious and sometimes flamboyant style had served him well. Watson's career had certainly been nurtured and encouraged by James Rowland Angell, and his development as a comparative psychologist had been heavily influenced by the functional psychology developed by Angell at Chicago. But Watson's decision to leave the University of Chicago indicated a determination to strike out on his own in more ways than one. In the fall of 1907, he had confided to Robert M. Yerkes that he had become convinced that it was "not up to the behavior men to say anything about consciousness."[83] To suggest that consciousness was not the essential subject matter of psychology flew directly in the face of accepted theory and practice. It was to be five years before Watson felt secure enough to issue his challenge publicly. During that time he developed the methodology and theory that became known as "behaviorism." As he prepared to leave for Johns Hopkins, he was equipped with a solid background in the study of animal behavior. Now he was ready, he wrote to Yerkes, "to get busy on the human side."[84] When Watson took up his duties at Johns Hopkins, it marked the establishment of experimental psychology there for the first time since G. Stanley Hall, the organizer of the profession, had left more than a generation before.

4

The Making of a Psychologist: 1908 to 1913

. . . I am anxious for Hopkins to come into its own . . . I am sure that we can make our institution a real power.

JOHN B. WATSON TO E. B. TITCHENER[1]

In 1903, James Mark Baldwin had been brought from Princeton to Johns Hopkins to build a department of philosophy and psychology. It had been his plan from the beginning to bring in someone to head a psychological laboratory and teach courses in experimental and comparative psychology. Watson was an extremely attractive choice for Baldwin. It was hoped that Watson would provide a strong link between the theoretical work being conducted in philosophy and psychology and the "more practical" work in physiology.[2]

For Baldwin, Baltimore was "dirty!—Southern in its ways. . . . and smelly in its alleys." But it was also "sincere and frank to the newcomer."[3] When Watson joined the Johns Hopkins faculty in the fall of 1908, he felt that the "whole tenor" of his life had changed. For the first time he "tasted freedom" in being able to work without supervision and to develop research along his lines of interest. He plunged into his work and soon became "lost . . . and happy."[4]

The psychology and philosophy department at Johns Hopkins had enjoyed a renaissance under Baldwin. By 1908, he had raised the department's budget to the fifth largest in the university. Baldwin was an influential psychologist and had a considerable impact on the development of the profession. In 1893, he had organized the Psychological Review Company with James McKeen Cattell. Designed to compete with G. Stanley Hall's *American Journal of Psychology*, this privately owned venture was established only a year after the creation of the American Psychological Association. Along with the *Psychological Review*, Cattell and Baldwin also published the *Psychological Index*, *Psychological Monographs*, and the *Psychological Bulletin*. These journals published papers covering the entire spectrum of psychological research. They also devoted a great deal of space to the professional activities of the American Psychological Association. In 1903 Cattell sold his interest in the company to Baldwin. When, in 1909, he was selected president of the Eleventh International Psychological Congress (to be held in 1913—the first scheduled to meet in America), Baldwin occupied a place of prominence and influence in psychology.[5]

Nevertheless, events of that year were to demonstrate the fragility of that position when codes of moral conduct were violated. Caught in a police raid on a Baltimore bordello, Baldwin succeeded in having his case quietly dismissed, and university officials hoped that would be the end of the matter. But when Baldwin was nominated to a position on the local school board, innuendos began to appear in the press, and the president and board of trustees of the university demanded his resignation. Baldwin, whether because of his own inability to face his colleagues or because of their failure to rally to his defense, isolated himself from the American psychological establishment. He moved to Mexico, where he spent several years, and eventually settled in Paris, where he was welcomed by French psychologists.[6]

Baldwin's misfortune was, ironically, a windfall for Watson, who did not hesitate to take advantage of his opportunity. Responsibility for the development of psychology at Johns Hopkins now rested entirely in his hands. He also assumed what he called the "fat and juicy job of running the *Psychological Review*." Thus, at thirty-one, Watson became the director of psychology at a major research institution and editor of a journal of considerable influence within his profession. Now he would have access to funding for his own research and a forum for the dissemination of his ideas. [7]

The responsibilities he assumed had their corresponding pressures. But despite bouts with insomnia and the burden of administrative and teaching duties, Watson continued to press forward with his research lest his hard-won professional standing erode. He had been criticized for concentrating his efforts on the study of animals to the neglect of human psychology, and he worried that a reputation based on animal work would be "ephemeral." Although currently in vogue, he wondered how long it would remain so. At last he resolved that he would work to make it a "respectful [sic] business."[8]

Watson expected no less from the members of his department. He complained about his colleagues and characterized those less willing than he to devote themselves to research as "lacking in grit." Surveying the potential of the department in light of his own expectations and in view of his ability to elicit support from the university administration, he was sure that as far as psychology was concerned, he could make Johns Hopkins "a real power."[9]

Watson seized every opportunity to consolidate his position. His motivation came not only from a desire for professional prestige and power but also from consideration of personal comfort and social status. In the fall of 1909, Robert Yerkes made an overture to Watson about the possibility of taking a "Chair of Experimental Education" at Harvard. Watson responded enthusiastically. He felt that such a program could be developed to the point where there would be "no competition" with similar programs in other institutions. But the price for Watson would be dear. Although he constantly complained about debts, he believed he had the best chance for advancement at Johns Hopkins. He hated Baltimore and was unhappy with the professional caliber of his colleagues at the university, but he felt that it would be a "strategical [sic] mistake professionally" to accept Yerkes's offer unless Harvard were to substantially improve his salary. [10] However, he was not above using the circumstances to improve his position at Johns Hopkins. Writing to university president Ira Remsen, Watson assured him that he was not likely to leave Hopkins for anything Harvard could offer—nevertheless, he felt he should be "frank" concerning his dissatisfactions. Regarding the question of salary, Watson boldly asked for a commitment to raise his income by a third over an unspecified period of time.[11] He was "determined to get a living wage," he confided to Yerkes, so that he could get out of debt and "live more completely" in his work. But he also looked for the time when money would provide the leisure and affluence to "join the golf crowd."[12]

Watson was contemptuous of those in his department who failed to meet his personal standards for achievement. He considered Edward Buchner, who directed the department's pedgogical wing, to be "an egregious ass" and a "high class janitor" hired to "coax these hayseed teachers to eat out of his universities [sic] hand."[13] Perhaps forgetting that he was once one of those "hayseed teachers," Watson complained that Buchner was not a "university man." Watson was concerned with the standing of the department among psychologists. Although such men as Buchner might be good instructors, he admitted, they were not "productive" factors in the sense that they did not add to the strength that could be seen "from the outside." Lest the work at Hopkins was to become a "negligible factor" in the scientific community, he warned Remsen, what was needed was no less than a thorough reorganization. Watson wanted an independent department of psychology, entirely separate from philosophy. He explained to Remsen that his own work had no connection with philosophy and that as far as he was concerned, he shared a "more intimate relation" with biology. Citing Cornell and the University of Chicago, Watson pointed out that Johns Hopkins's competitors had a head start in this regard.[14] Watson wanted psychology to be established on an equal footing with the rest of the natural sciences, and he believed that its acceptance by the scientific establishment depended upon its ability to produce results.

Watson wanted to demonstrate that psychology was a true experimental science, based upon a sound methodology but capable of producing knowledge with practical applications. He realized, as had Hall and Cattell a generation earlier, that in order to grab the attention of university administrators and foundation directors, he must arouse the interest of a broad audience. Writing in 1910 for *Harper's* on "the new science of animal behavior," he argued for a redefinition of psychology based upon what he considered to be the "progress" of the discipline. Just as "speculative or metaphysical" philosophy had been replaced in the previous generation with a "new psychology" that attempted to map the structure of consciousness, the methods and assumptions of experimental psychology were being challenged by a new science, which Watson called the science of animal behavior. Experimental psychologists may have borrowed the form of scientific investigation from the physical and biological sciences, Watson argued, but since their experiments involved the method of introspection, wherein the investigator observed his or her own conscious activity, there could be no control in any true scientific sense. Furthermore, a true science of mind would have to assume the evolution of consciousness, just as biology was founded upon the evolution of the structure of a species's physical characteristics. If psychology could not accommodate the study of animals, then its assumptions and methodology must change. Watson believed that if a correlation between behavior and mental activity could be found, psychology could become a true science capable of encompassing the whole evolutionary scale. Watson's new science was centered upon the observation of animal behavior. By allying himself with the biological sciences, Watson hoped to spearhead a move that would change the fundamental emphasis in psychological research. Above all, Watson pointed out to the readers of *Harper's*, a new approach to the study of psychology would have enormous practical benefits. Watson believed that coming to an understanding of the learning process was the central problem in human psychology. Assuming that learning takes place by trial and error in both man and beast, Watson argued that an investigation of that phenomenon by controlled experiments on animals could lead not only to an increased efficiency in human learning but ultimately to control over the learning process itself.[15]

Watson cautioned his colleagues not to take his popular writings too seriously. They were "pot-boilers," he explained, written merely "for the money." Watson's apologetic tone was, perhaps, intended to forestall criticism from some of his more scrupulous colleagues. Though he may have disavowed his rhetorical excesses, he was deadly serious about the effect he hoped his articles would have. Watson intended for them not only to "excite popular interest" in psychology but also to enhance his own position at Johns Hopkins. He complained to Robert Yerkes that he lived in a community that "practically never heard of psychology." Moreover, he felt that his department's allocations depended on the extent to which he made the administration and the board of trustees "take notice" of psychology. Watson was proud of the effectiveness of his articles. He wrote to Yerkes that members of the board of trustees, including the president himself, had personally sent their congratulations. Now he would be able to get from them, he boasted, "just about what I want."[16]

Watson was confident that he could secure Ira Remsen's support for holding the Eleventh International Psychological Congress at Johns Hopkins.[17] Representing the host institution for the first international meeting of psychologists to be held in America would have been an impressive addition to his growing collection of achievements. But the Congress was ill-fated from the beginning, and the failure of American psychologists to bring it about is illustrative of conditions within the struggling profession at that time.

James Mark Baldwin had been chosen president of the Congress, and American psychologists were put in an embarassing position when he was forced to resign from Johns Hopkins. After Baldwin's removal from the presidency of the Congress, his colleagues were confronted with the decision of whether to continue with plans for the Congress and, if so, who would be its new president. Arguments in favor of continuing with the Congress involved concerns of both national pride and professional prestige. Writing to William James, James McKeen Cattell maintained that psychology's standing in other countries and among other sciences in America must not be allowed to deteriorate. But aside from avoiding negative repercussions, Cattell believed that holding the Congress would be extremely useful in advancing the interests of psychology in America. The public-relations possibilities of an international congress appealed to Cattell; besides, psychology had to keep its professional activities up to date. Amongs the sciences, he pointed out, "the international congresses are here to stay" and psychologists might as well accept their "fate."[18] E. B. Titchener of Cornell echoed Cattell's sentiments about the necessity of saving face. For him, nothing less was at stake than "our national dignity and honour."[19]

Upon Baldwin's resignation, the remaining officers of the congress were Titchener and Cattell as vice-presidents and Watson as secretary. These men represented three factions among American psychologists. Titchener, alone in America, represented the Wundtian structuralist school in an extreme and unmodified form. In 1904 he had challenged the American Psychological Association with a bid to form a rival organization: the American Society for the Advancement of Experimental Psychology. Psychologists, for the most part, declined to affiliate with Titchener's group and rallied to support the Amercian Psychological Associaton. Thus the only serious challenge to the APA's hegemony over the profession was easily resisted. Titchener's society continued to exist, however, in the form of small, informal meetings to which a select group of psychologists were invited to discuss their research.[20] Cattell, head of the influential Department of Psychology at Columbia University, had long been a promoter of the

APA and an advocate of applied psychology. Using his position as editor of *Science, Popular Science Monthly,* and *School and Society* to promote psychology, he was a powerful influence within the APA. Watson represented the younger generation of psychologists, mostly trained in America, who were restless under the leadership of those who had founded the profession. Although Watson longed to chart his own course, he was careful to maintain ties with both of the other factions. Watson was long a member of Titchener's experimental group, despite his opposition to Titchener's doctrinaire structuralism, and he took pains (although his impetuosity often got the better of him) to cultivate Cattell's goodwill as well.

The fortunes of the Congress mirrored, to a large extent, the conflicts among these three groups as each maneuvered to dominate the proceedings. Titchener was willing to support Cattell's nomination of G. Stanley Hall as president, but Watson would have none of it. Watson considered Hall to be a stalwart of the old guard of the APA and out of touch with current research.[21] Watson wanted to push Titchener into the office in a move against Cattell and Hall, but Cattell opposed him on the grounds that Titchener was an Englishman and, more important, was not even a member of the APA.[22] Referring sarcastically to Watson as a man of "peace and policy," Cattell complained to William James of Watson's contentiousness. In a typical display of histrionics, Watson took elaborate offense at Cattell's barb. Pretentiously claiming to Yerkes that he was brought up as a "gentleman and among gentlemen," he did not take lightly what he alleged to be his "first insult." Looking forward to the day when psychologists of his generation would "grow into the saddles," he felt that they would set a much "better example to those who are to follow." Watson was anxious to hasten that day, for he felt that "it would be a fortunate thing for psychology if Cattell were to be called upon to edit his *Science* and *American Men of Science* with its 'list' [ranking "prominent men of science"—among which Watson was not included] up in heaven."[23]

In order to "pour oil on the troubled waves" rocking the Congress, William James, weakened and in ill health, tried to reconcile the conflicts between the contending factions. Having been urged by Cattell and others to accept the presidency in order to save the Congress, James, in a letter to Watson, explained that he was concerned about the disgrace that would befall American psychology should the congress fail to be held. Although contrary to his own inclination, James agreed to accept the nomination for president, in the interest of preserving peace, providing his position would be honorific. With James as titular head of the Congress, the balance of power was preserved among the competing groups, with no one gaining or losing advantage.[24] But James's death in the summer of 1910 disrupted the uneasy coalition. Despite subsequent attempts to salvage the Congress, failure to agree on an acceptable president plus dire reports from Hugo Münsterberg on European psychologists' lack of interest, led to final abandonment of the project.[25] Watson hoped that at last it would "rest in peace."[26] The inability of psychologists to arrive at a consensus on a president for the Congress reflected a crisis in leadership and focus for American psychology. The founders of the profession were passing away. Their successors were being challenged by those of Watson's generation who were impatient with the methodological limitations of the experimentalists. What was lacking, they complained, was a unified theory that could serve as a rallying point for their aspirations. As he began to articulate these concerns, Watson found a ready consensus.[27]

Psychologists found themselves still struggling for acceptance from the scientific community. The strong personalities that had shaped the profession were steadily diminishing in number and influence, and their decline created a vacuum in leaderhip as well as a lack of clear definition for a profession still caught in a crisis of identity. In 1908, Watson had complained to Titchener that "Münsterberg no longer does experimental work, Cattell no longer does experimental work, Angell is not publishing, Judd will go to Chicago and get swamped in administrative work and where will our big laboratories draw their inspiration from? Really if we are to keep our steps directed onward and not backward something must happen." Watson regretted the lack of decisive leadership within the profession. He viewed the failure of the Congress as an indication of stagnation within American psychology. "I get pretty blue sometimes," he wrote, "when I look at the men who ought to [be] patriarchs in the experimental work and see that they are giving up the fight and going into—God knows what."[28] Watson emerged from the struggle over the Congress as a strong advocate of overhauling the profession. As he continued to address this issue, he became an increasingly influential voice among psychologists.

In 1910, Watson's position was further enhanced when he established, with Robert M. Yerkes, the *Journal of Animal Behavior*. Watson and Yerkes wanted to provide a vehicle for the dissemination of their ideas as well as an outlet for research that supported their conception of a behavioral science. While protesting to Yerkes that he was not interested in creating a journal for the purpose of establishing a base of personal "power" (an accusation that he made against influential figures like Cattell and Angell), Watson nevertheless agreed with Yerkes that a prime reason for the venture was to influence the direction of psychology by "helping to shape the course of publications." Watson wanted to move psychology closer to the biological sciences, and the journal was designed to bring about a new alliance between psychology, physiology, zoology, and anatomy.[29]

Watson characteristically drove himself at a frantic pace that often brought him to the brink of nervous exhaustion. He would seek and accept responsibilities only to complain bitterly of overwork. In addition to his departmental duties, his regular teaching load, and his own research, he also put his energies into planning for the psychological congress, editing the *Journal of Animal Behavior*, and co-editing the *Psychological Bulletin*. During this time Watson also combined a summer teaching schedule with additional research in the Dry Tortugas. And he carried other burdens. Despite a successful effort to extract a substantial salary increase from Remsen with the threat of another job offer,[30] Watson continued to bemoan the "incubus" of his debts. Perhaps his continual concern with financial debt reflected fears of depleting what he considered to be scarce emotional resources. The pressures of career and family must have had their repercussions. But Watson, always one to objectify emotional tensions, never related his family's stresses to his own. In a letter to Yerkes, Watson fretted over what he called his son's unbalanced nervous system. He then related the story of his attack of "nervous prostration" while a student at the University of Chicago and confessed that he now felt himself to be "on the verge of a breakdown." Lectures were delivered "with great difficulty," and he had to watch himself carefully.[31] He constantly struggled with the fear of losing control, and he usually reacted by working even harder.

Watson divided his summers between research at Bird Key in the Tortugas Islands during the nesting season and teaching summer sessions for Cattell at Columbia or for Angell at the University of

Chicago. The research in the Tortugas was sponsored by the Carnegie Institution of Washington and coordinated by Alfred G. Mayer. Mayer felt that Watson's research was among the "most important we have been privileged to aid" and offered to pay half of Watson's stipend personally so that he could afford to spend the entire season on the project. Writing to R. S. Woodward, the president of the Carnegie Institution, Mayer complained of the "summer teaching trap" that caught so many scientists struggling to make a living and described the situation as a "most serious menace to the prospects of research." Finding the problem to be growing yearly "more intolerable," he vowed to begin a personal crusade and felt that he "*must* protest or research will be crushed!" Although able to get regular increases in salary from Johns Hopkins (usually in excess of his peers at other institutions), Watson gave Mayer the impression of one hovering on the brink of financial disaster. Mayer believed Watson to be "one of the ablest, most courageous and cheerful workers in his field our country has developed" and felt a "disagreeable creepiness" when he considered that Watson's research was being held up because of the necessity of repaying debts. Accordingly, he set out to "rescue" Watson from the "summer school pitfall" in order that he might be "saved" for research. Persuaded by Mayer's arguments, Woodward offered Watson a stipend that would pay for three summers' work.[32]

Watson's research in the Tortugas involved studying the migrating and nesting habits of a species of terns. There, on a barren strip of sand, Watson marooned himself during the heat of the summer with five thousand birds for company.[33] Watson was interested in collecting data on instinctual behavior. His dissatisfaction with the Darwinian conception of instinct—which called for a belief in the "'fitness' or 'adaptiveness'" of all instinctive activity—led Watson to gather new data by "experimentally controlling the process of evolution." The central question was: To what extent are fixed modes of responding inherited, and to what extent are organisms equipped with "plastic forms of activity" that require shaping by training or instruction?

Clearly, Watson believed that the acquisition of this knowledge would lead to practical applications. In an article in *Harper's*, he explained that it would provide the means of encouraging desirable tendencies and suppressing undesirable ones. It was the "practical" reason, he argued, that justified investing so much time on the study of animal instincts.[34] Watson was anxious to test his ideas. In a letter to Hugo Münsterberg, he asked about the feasibility of establishing a "vocational bureau" supported by business and professional groups.[35] Watson wanted to explore the reasons why so many promising men, owing to what he called a "lack of balance in the individual," are put at a disadvantage in competing for jobs. Watson felt that vocational training could identify and eliminate tendencies that hindered successful careers and could help develop traits that were consonant with good work habits. Such a system would, he believed, not only "spare us many a neuropath and many a criminal" but also produce a "higher level of efficiency"—even if it would not lead to the production of genius. Calling for a a new type of professional educator, Watson outlined a system whereby a teacher trained in psychological methods would bring up a "squad" of children from the primary grades through high school. These instructors could shape behavior and provide data to psychologists for the preparation of general vocational tests. Although Watson felt that such knowledge was lacking at present, he was confident that the next few years would bring dramatic results. "Now that psychologists are breaking away from academic tradition," he wrote, they "are willing to admit that psychology has practical outlets."[36]

But Watson still felt constrained by lack of support for applied psychology. He had struggled to sever the link between psychology and philosophy and had established an independent department of psychology at Johns Hopkins.[37] He also had sought ways to demonstrate psychology's application to more practical concerns. Educators, with their proven interest in applying psychological techniques to pedagogy, provided a broad constituency to psychologists who wished to extend their influence. Watson had developed "a deep interest in the experimental phases of education and in applied psychology generally." But he became alarmed when he learned that courses in applied psychology were being offered within the Education, rather than Psychology Department, at Johns Hopkins. Watson saw this as an attempt to usurp the prerogatives of psychologists. The issue was important enough for Watson to enlist the support of John Dewey, E. L. Thorndike, and James Rowland Angell. Backed by Adolf Meyer and Arthur O. Lovejoy at Johns Hopkins, he presented his case to university president Remsen. Watson suggested that the university secure a psychologist who would offer courses in "experimental pedagogy, child psychology and possibly applied psychology" under the auspices of the psychological laboratory. For psychology to advance as a science, it had to achieve and maintain a secure professional niche within academia.[38]

Other psychologists echoed Watson's concerns about the professional standing of psychology. In 1912, a report on the status of psychology appeared in the *American Journal of Psychology*. Although some gains were reported in the growth of independent departments, psychology did not compare well with other academic disciplines. It was the university through which the profession sought to establish itself, and psychology aspired to the recognition and security that had been achieved by the physical sciences. In comparison with other academic fields, the report continued, psychology placed last in number of professors, hours of instruction, and expenditures. Complaints from psychologists generally concerned the dependence of psychology upon philosophy. The survey found that philosophers were criticized for their unscientific and nonempirical methods and for a "'philosophizing tendency' which works havoc with the empirical approach attempted by modern psychology." Where there was harmony reported between the two fields, the "psychology" in question turned out to be the "non-empirical generalizing variety current before 1880." Most psychologists felt that if an independent department was not possible, an academic relationship with the biological sciences would be more desirable than with philosophy. A widely held opinion was that philosophy was a millstone around the neck of the younger discipline. The report ended by suggesting that pure sciences that ignore applications of their research were usually slow in academic growth. It also warned that the introspective method itself, which was "peculiar to the psychologist may offer a hindrance to the ready acceptance of the discipline." Suspect in scientific circles and poorly understood even by psychologists, introspection was looked upon by a growing number of psychologists as a barrier to professional recognition.[39]

Introspection had long been a major reason for Watson's dissatisfaction with the mainstream of experimental psychology. Only a year after coming to Johns Hopkins, Watson had struggled with his own conception of psychology. He read Darwin for the first time while developing an outline for a proposed textbook. Exasperated, he wrote to Yerkes:

Damn Darwin. The Neo-Darwinians and Neo-Lamarckians, etc. are in a worse hole than psychologists—! I am terribly at sea as to finding a proper place and scope for psychology. What are our simple presuppositions and what are we good for? . . . I have come out of this—one chapter will have Behavior a biological problem— the scientific determination of modes of behavior and the *modus operandi* of behavior—a part of the problem of natural selection—the second the psychological implications in modes of behavior. My interests are all in the first where an objective standard of determination is possible and where interpretation takes the line of the *importance* of the *observed facts*—for the theory of selection—facts— and interpretation possible without mentioning consciousness or deviating from a (wide) biological point of view. What is there left? Am I a physiologist? Or am I just a mongrel? I don't know how to get on.[40]

Watson had great difficulty reconciling prevailing attitudes among psychologists with current scientific practices. Watson's solution was to define behavior as a biological problem while ignoring consciousness. It was the only way to satisfy demands for an "objective standard of determination." Yerkes confided to E. B. Titchener that he and Watson were "just on the point of becoming psychologists or turning from it into physical science forever." He felt that neither he nor Watson were given much encouragement by experimental psychologists and was discouraged that little effort was made to bring them "into the fold." Yerkes hoped that Titchener would make some effort to modify Watson's extreme position.[41] Watson, however, had made up his mind. He considered Titchener to be an embarrassment. He "takes us away," Watson complained to Yerkes, "from the good graces of the physicist."[42]

By early 1910, Watson had further clarified his position, although it was to be three years before he was to make his views known to the psychological community. Frustrated by the low status of psychologists among the natural sciences, Watson protested to Yerkes:

I am a physiologist and I go so far as to say that I would remodel psychology as we now have it (human) and reconstruct our attitude with reference to the whole matter of consciousness. I don't believe the psychologist is studying consciousness any more than we are and I am willing to say that consciousness is merely a tool, a fundamental assumption with which the chemist works, the physiologist and every one else who observes. All of our sensory work, memory work, attention, etc. are part of definite modes of behavior. I have thought of writing . . . just what I think of the work being done in human experimental psychology. It lacks an all embracing scheme in which all the smaller pieces may find their place. It has no big problems. Every little piece of work which comes out is an unrelated unit. This might all be changed if we would take a simpler, behavior view of life and make adjustment the key note. But I fear to do it now because my place here is not ready for it. My thesis developed as I long to develop it would certainly separate me from the psychologists—Titchener would cast me off and I fear Angell would do likewise.[43]

By 1913 Watson felt more secure as a psychologist and less reticent about challenging the leadership of the profession. He was not willing to follow Yerkes's advice to "let psychology go its own gait" and objected to his suggestion to merely "call behavior physiology or biology and leave psychology to the introspectionists." Watson believed that he knew how to make psychology a "desirable field for work." He was "not willing to turn psychology over to Titchener and his school."[44] By redefining the field of psychology and systematizing it, he hoped to align the profession with the older branches of science. Accordingly, he arranged to present his views in a carefully planned series of lectures and articles that were to establish his reputation as the creator and standard-bearer of behaviorism.

5

Crying in the Wilderness: The Advent of Behaviorism, 1913 to 1917

I get rather disgusted sometimes with trying to make the human character amenable to law.

JOHN B. WATSON TO ROBERT M. YERKES[1]

To the educated middle classes, the spring of 1913 seemed to vibrate with the shock of new ideas and art forms that threatened to topple the ornate palaces of official Victorian culture. Americans read of riots in Paris due to the performance of Stravinsky's *The Rite of Spring*. In February and March of that year the Sixty-Ninth Regiment Armory on New York's Lexington Avenue was crowded with those who came to gape at a collection of paintings by a group of artists who challenged traditional stylistic conventions. The Armory Show of 1913 drew almost one hundred thousand visitors, who responded to the exhibit with a mixture of awe and outrage. But it was a critical and financial success—due in no small measure to the carefully staged showmanship of the exhibit's organizers. The show represented the attempt of artists to reflect life in the twentieth century, free of the philosophical and stylistic restraints of the past. The exhibit may have seemed revolutionary to an astounded public, but it was an indication of the self-confidence of those who had for some time considered themselves to be part of an *avant-garde* and now felt secure enough in their position to confront the bourgeois world. The message of the Armory Show was not a revolutionary manifesto, but rather a pronouncement to a startled populace that a coup had already taken place. The modern era had arrived.[2]

Ninety blocks to the north on Morningside Heights, John B. Watson was making a proclamation of his own, with a series of lectures at Columbia University which sent reverberations throughout the psychological community. Watson's lectures coincided with those of French philosopher Henri Bergson, whose blend of science and spiritualism appealed to the progressive sensibilities of New York's well-to-do. Broadway was thronged with those making their way uptown to the Columbia campus. Though Bergson was the major attraction, Watson held his own against the competition. The lectures were crowded, and Watson modestly expressed surprise at the size of the audience. He failed to see how they found "so much to interest them" he demurred in a letter to Robert Yerkes, "since [he] had not planned to make the lectures popular."[3] But Watson characteristically protested too much about his popular success. His remarks to Yerkes scarcely concealed his glee at his personal triumph. His first lecture, "Psychology as the Behaviorist Views It," was carefully calculated to draw the attention of psychologists.[4] By styling himself as "the behaviorist," Watson was staking out a position for himself not only as a critic of current psychological theory and method but also as the champion of a distinct and separate approach to experimental psychology. Watson coined the term *behaviorism* (which he italicized for emphasis in the published version of his address) to distinguish his brand of psychology from that of the structuralists and functionalists. The unveiling of behaviorism was the first public articulation of convictions that Watson had held for some time. He used the occasion not only to challenge psychologists' fundamental assumptions but also to call for radical measures to prepare the way for a science of behavior that would fulfill what he believed to be the true promise of psychology.

Watson began his lecture with a sweeping declaration that sought to establish, once and for all, the precise nature and parameters of the science of psychology. Psychology, he insisted, was a "purely objective experimental branch of natural science," with its "theoretical goal" being nothing less than the "prediction and control of behavior." By implication, Watson was quite clear about what, in his opinion, psychology was *not*. It was not a stepchild of philosophy. Speculations about the nature of mind that could not be tested in the laboratory had no place in an "experimental branch of natural science." Furthermore, those who conducted experiments that could not be controlled or who insisted on assumptions that could not be verified were hardly scientists and certainly not psychologists. He argued that introspection should be ruled out as a method of investigation since the "scientific value" of psychology could not depend upon the capacity of its data to be interpreted in terms of such a vague and imprecise concept as human consciousness. Like biology, Watson argued, psychology should be able to discover natural laws that applied to all living organisms. He wanted to remove the last barrier to a naturalistic concept of the evolution of intelligence, for, as he put it, he recognized "no dividing line between man and brute." This was a remarkable statement. It was at variance not only with popular notions of human development but also with practically every scientific fact. It was Darwinism taken to its logical limits, for it implied that man's appearance was but a mere chimera in the evolutionary scale. But for Watson, it was an essential position if behaviorism was to have a secure theoretical foundation. His objective was to describe a "unitary scheme of animal response." But Watson did not stop there. It was not enough for psychology to have a theoretical and methodological footing that would enable it to make an uncontested contribution to scientific knowledge. Watson had a grand vision of its applications. The fruits of psychological research would have an enormous impact on the development of humanity, he believed, for, as he emphasized at the beginning of his lecture, the ultimate goal of psychology should be not only to be able to predict behavior but to *control* it.[5]

The forcefulness with which Watson made his points was partly engendered by his frustration with the constraints under which he had felt compelled to work. He had, for some time, felt hard-pressed to explain his work in animal, or comparative, psychology to psychologists who evaluated all data in terms of their contribution to the understanding of human consciousness. Watson confessed that he was often "embarrassed" by this situation. Although in his own research questions regarding animal behavior had been answered to his satisfaction, he felt pressured by the prevailing definition of psychology to draw conclusions as to the possible mental states of animals. Watson was no longer willing to work under these "false pretenses." He demanded that "either psychology must change its viewpoint so as to take in facts of behavior, whether or not they have bearings upon the problems of 'consciousness'; or else," he threatened, "behavior must stand alone as a wholly separate and independent science."[6]

Watson maintained that the assumption of the presence or absence of consciousness had absolutely no effect upon what he called the "mode of experimental attack" in the investigation of behavior.[7] But aside from the irrelevance of consciousness, the larger issue for Watson was that such "esoteric" concepts had held psychology back from a full acceptance by the scientific community. Claiming that psychology had "failed signally" to take its place as "an undisputed natural science," Watson blamed the use of the introspective method and the underlying assumption of the existence of states of consciousness. It was a method that blamed the observer rather than the experimental setting if results were not obtained. In physics and chemistry, Watson argued, the reverse was true: The "attack" was always made upon the experimental conditions. For Watson, the time had come "when psychology must discard all references to consciousness: when it need no longer delude itself into thinking that it is making mental states the object of observation.[8]

The functional school of psychology had not gone far enough in this regard to suit Watson. Indeed, he saw little difference in practice between structural and functional approaches and found himself confused by the terminology of both schools. Watson felt that behaviorism was "the only consistent and logical functionalism." It avoided such issues as the "mind–body problem" (a subject ignored, Watson noted, by other branches of science), as well as the use of such terms as "consciousness, mental states, mind, content, introspectively verifiable, imagery and the like." Watson felt that a new system of behavioral psychology could be written in terms of "stimulus and response . . . , habit formation" and "habit integrations." This psychology would be based on "the observable fact that organisms, man and animal alike do adjust themselves to their environment by means of hereditary and habit equipment." He envisioned a system of psychology in which "given the response the stimuli can be predicted; given the stimuli the response can be predicted."[9] Watson was clear about the "final reason" for the development of behaviorism: It would enable him, as he put it, "to learn general and particular methods by which I may control behavior."[10]

Watson's emphasis on control was central to his concept of

behaviorism. For apart from his theoretical and methodological objections to structuralism and functionalism, Watson had long been frustrated with the difficulty of using those approaches to demonstrate the practical applications of psychology. It must not be overlooked that Watson's key argument for the acceptance of behaviorism was that it could produce techniques with spectacular socal applications. "If psychology would follow the plan I suggest," he wrote, "the educator, the physician, the jurist and the business man could utilize our data in a practical way, as soon as we are able, experimentally to obtain them." Watson noted that the fields of experimental pedagogy, the psychology of drugs, the psychology of advertising, legal psychology, the psychology of tests, and psychopathology were all "vigorous growths." The fact that such fields were the branches of psychology most removed from traditional psychology and the least dependent on introspection convinced him "that the behaviorist's position is a defensible one." These areas, he pointed out, were "truly scientific" in their own right. They were not dependent on previously acquired knowledge, but provided new discoveries "in search of broad generalizations which will lead to the control of human behavior."[11] Behaviorism, Watson argued, would, in effect, provide the tools with which psychologists would become social engineers.

Watson's lecture at Columbia was designed both to admonish and to inspire. By pursuing the study of an elusive and unverifiable consciousness, he contended, psychologists had become lost in a wilderness inhabited by philosophers, pseudo-scientists, and other wandering souls. Watson proclaimed behaviorism to be the means by which psychology could achieve the status of a rigorous experimental science that would discover and apply the laws governing human behavior. He envisioned a promised land of which psychologists had long dreamed. Behaviorism, Watson reasoned, was not an abandonment of fundamental psychological principles but a radical shift in method that offered a sure path toward the fulfillment of psychology's true destiny. Watson addressed issues that had been of concern to psychologists for some time. He was certainly not the first psychologist to criticize the use of the introspective method. James McKeen Cattell, a devoted partisan of applied psychology, had made a strong argument against introspection as early as 1904 (at a meeting attended by Watson).[12] In 1910, Watson's former professor, James Rowland Angell, had predicted that the term "consciousness" would fall into the same category of disrepute into which the word "soul" had been cast. Angell, however, had not meant to deny that the phenomenon of consciousness existed; but he had clearly perceived what he described as a "shift of psychological interest toward those phases of [inquiry] for which some term like behavior affords a more useful clue."[13] Angell, not unaware of Watson's leanings, no doubt had him in mind when he made the prediction. Although Watson did not originate the "shift" that Angell described, he articulated in a clear, forceful, and dramatic way the concerns of many in the profession.

The sensational manner in which Watson made his case had as much to do with the impact of Watson's message as did the content of his argument. He had drawn a sharp line between what he considered to be the forces of reaction and the champions of progress in his profession and thereby defined the terms of debate regarding the scope and purpose of psychology. Watson's declaration provoked a widespread response from social scientists and psychologists. If the reaction from psychologists was mixed, it reflected a profession that was divided against itself. Although many applauded his critique of experimental methodology, few were willing to go along with the wholesale abandonment of consciousness as an area of investigation. But speaking for a growing number of psychologists, Watson drew together themes that had long been undercurrents in the profession and provided a synthesis that held out the promise of a "useful" science.[14]

The initial responses to the issues that Watson raised in his Columbia lectures ranged from cautious optimism on the part of most psychologists to wholesale condemnation from the few remaining partisans of structuralism. Those whose progressive sensibilities favored the rationalization of society through science were the most enthusiastic. John Dewey announced that he was a "well-wisher" of behaviorism. Dewey welcomed the "radical" implications of behavioral methodology. He characterized the "orthodox psychological tradition" as having arisen not "within the actual pursuit of specific inquiries into matters of fact, but within the philosophies of Locke and Descartes." The adoption of behaviorism would not mean merely the initiation of new ways of dealing with old problems but the "relegation of [those] problems to the attic in which are kept the relics of former intellectual bad taste." But questions of taste aside, Dewey feared that despite behaviorism's promise, it had within itself the seeds of its own destruction. "In so far as behaviorists tend to ignore the social qualities of behavior," he warned,

> they are perpetuating exactly the tradition against which they are nominally protesting. To conceive behavior exclusively in terms of the changes going on within the organism physically separate in space from other organisms is to continue that conception of mind which professor [R. B.] Perry has well termed "subcutaneous."[15]

Mary Whiton Calkins, who had been a student of William James and had founded the psychology department at Wellesley, also criticized the one-dimensional, mechanical character of behaviorism and maintained that a psychology of behavior must take into account both the social and biological aspects of behaving. But she supported Watson's critique of the "undue abstractness" of psychology as well as his efforts to make psychology more useful to "those concerned with problems of life." She was not, however, willing to abandon introspection. She was convinced that the introspective method, aside from its primary function as a technique for investigating mental phenomena, was useful in ordering and modifying the behavior of those who used it. It was, she argued, a means of acquiring mental discipline. "Drill" in the analysis of consciousness, Calkins believed, provided practice in discovering some order "within the bewildering complex of experience" and "offers valuable training to young students." Like Dewey, James, and Watson, Calkins believed that the test of a useful psychology was the extent to which it enabled human beings to channel primitive impulses into social usefulness. Her disagreement with Watson was that behaviorism seemed to rule out the possibility of inner regulation through self-discipline and to substitute the primacy of natural law over that of individual choice. Those whose sensibilities had been shaped by late Victorian culture could welcome Watson's vision of the social applications of psychotechnology as a step in the march of progress as long as behaviorism did not stray from the path of moral order.[16]

The most forceful critique of Watson's behaviorism came from E. B. Titchener, who considered himself to be the lonely champion of Wundtian structuralism in America. Titchener perceived a drift toward applied psychology, and his was one of the few voices raised in opposition to it. His fundamental criticism of behaviorism was that its "practical goal of the control of behavior" gave it "the stamp of technology." In Titchener's opinion, "to exchange a science for a technology" was "out of the question."[17] But this was precisely what Watson was asking. If his demand seemed radical it was because it required a redefinition of the scope of psychology in terms of changing perceptions of scientific value.

Several years later, Watson explained to Robert Yerkes the process that led to his decision:

> I came up through philosophy into introspective psychology, ran the laboratory at Chicago for years, doing both my animal work and looking after the problems and teaching straight introspective psychology. I had to put away all outside thought and fight to make introspective psychology scientific, and I have had many rows and arguments with biologists and others trying to make my points. The first few years at Hopkins I had to do the same thing, running both the human laboratory and the animal. Finally my stomach would stand no more and I took the plunge I did in 1912.[18]

Behaviorism grew out of Watson's desire to make psychology more useful and therefore more scientifically acceptable. Psychology had been founded, in the nineteenth century, with the hope of establishing an exact science of human nature by applying scientific methods developed by the natural sciences to the study of human beings. Confidence in this possibility came from the integration of mankind with the rest of the natural world made possible by Darwinian evolutionary theory—and from the positivistic character of nineteenth-century science, which emphasized that conclusions could be drawn only from data verified by observation. One result of this emphasis on positivism was that the prestige formerly afforded to scientific *theories* came to be given to scientific *methods*. The successful model for the development of a science was physics. Psychologists tried as much as possible to conform to that model in an effort to raise their discipline's status within the scientific community. Behaviorism was a logical development of this trend in that it sought to make the procedure of investigation in psychology as close as possible to that in physics. Ironically, Watson and his colleagues, like most of the scientific community, were ignorant of the revolution that was transforming physics. Already in 1905, an obsure patent-office clerk named

Albert Einstein had published the paper that would turn Newtonian physics to dust and rejuvenate the status of theoreticians. It was the physics of Ernst Mach, not Einstein, that provided the model for scientific achievement. Mach, who craved facts rather than theories, believed the basis of all scientific knowledge to be sense experience and placed great importance upon the scientist's ability to predict. For psychologists, the achievements of nineteenth-century physics had been impressive, and they were determined to follow its example.[19]

Behaviorism, as Watson initially presented it, was a declaration of faith. It was based on the belief that a specific methodology *could* transform psychology into what he considered to be a genuine science. All that Watson required of the faithful was the wholesale rejection of competing methodologies. By identifying with the positivistic orientation that characterized the development of the natural sciences toward the end of the nineteenth century, behaviorism at the very outset erected two powerful ideological defenses against criticism: (1) any nonpositivistic position was unverifiable and therefore unscientific and (2) positivism had no central doctrine that could be scientifically challenged. Any empirically verifiable statement was consistent with it. It was totally pluralistic in that the empirical findings of any challenging system could be accommodated.[20]

Behaviorism, as presented by Watson in 1913, was not a well-defined theoretical position. His work on the conditioned reflex and the development of a behavioristic learning theory came later. The significance of Watson's initial presentation of behaviorism was that it helped to crystallize fundamental issues that had created a crisis in psychology.[21] Psychologists who relied upon introspection as a methodology were not accepted into the professional scientific community, where such concepts as "mind" and "consciousness" were considered to be as empirically unverifiable as "soul." Watson resolved this dilemma by restructuring the framework of psychological investigation in accord with current scientific assumptions. Behaviorism was pragmatic in the sense that it insisted that the proper study of psychology was not mind but behavior. In short, Watson provided psychology with a theory and a methodology that satisfied the contemporary requirements for *being* a science. But behaviorism also satisfied the contemporary requirements for the *uses* of science, that is, the prediction and control of natural phenomena (in this case, human behavior) in the interests of efficiency, order, and progress.

In 1914 Watson published *Behavior: An Introduction to Comparative Psychology*, an introductory textbook he hoped would also interest the general reader.[22] In this way, he hoped, behaviorist methodology could be institutionalized in the classroom and laboratory, as well as brought before a wider audience. Watson's hopes were not unfounded, for in his argument for the adoption of behaviorism he addressed issues of vital concern to an emerging corporate society.

Behaviorism appeared at a time of national self-examination. The energies released by unparalled industrial growth threatened to rip apart a social fabric that seemed no longer able to accommodate the demands of urban expansion and increasing productivity. Watson's voice was part of a chorus of social critics and thinkers who offered solutions to this crisis of the social order. Walter Lippmann's *Drift and Mastery*, published in the same year as Watson's textbook, provides a glimpse into the contemporary issues that formed the context for the emergence of behaviorism. Lippmann saw a breakdown of order as a "nation of villagers" found themselves in a society that had rapidly become urbanized and industrialized. The individual and social forces released by changes in the social structure should not be passively accepted, repressed, or merely contained, argued Lippmann, but viewed as energy to be directed and channeled. Purpose was to be substituted for tradition, and efficiency rather than an appeal to authority was to be the measure of value. Science, for Lippmann, was to become a new religion, and the scientific method was to create a new binding faith for its practitioners. Lippmann explained that the scientific method had given man control of "progress" in providing an organization for the deliberate invention of technology. He saw the development of an "infinitely greater control of human invention" in sciences that were "learning to control the inventor."[23]

Psychology in particular was the discipline to which Lippmann looked to open up "greater possibilities" for the "conscious control of scientific progress." Psychology would be especially useful, he argued, to the new class of professional managers who were replacing the "profiteer" in business. These men now had to maintain contact with researchers in the factory laboratories and "deal with huge masses" of workers "becoming everyday more articulate." They had to consider the training public schools provided for the workforce and take into account such varied subjects as "the psychology of races," the credit structure, and international relations. They were also responsible for making the production and consumption process more efficient. Industrial education could make labor more productive, or, failing that, the inefficient worker could be replaced by a machine. Lippmann observed that it was "no accident" that universities had begun to create graduate professional schools to supply the specialists demanded by the expanding economy. The nineteenth-century virtues of hard work and thrift no longer assured either success or competence. "The universities," according to Lippmann, were "supplying a demand" created by "big business."[24]

Watson and Lippmann were addressing themselves to identical issues. Behaviorism called for the same faith in the scientific method to usher in a new era of progress in which the goal was "prediction and control" of behavior. Watson offered the services of psychology to the very class of managers whom Lippmann had characterized and held out the possibility of new professional roles for psychologists themselves. Both Watson and Lippmann displayed a "progressive" faith in a radical environmentalism that invested man with the ability to shape his own world. Human beings were molded by their environment, and the fate of mankind depended on a willingness to bring the process of social change under control.[25]

John Dewey was also attracted to the broad concept of behaviorism for precisely the same reason. Not only had he supported Watson's attack on introspective psychology, but he further argued that the most significant characteristic of behaviorism was its substitution of "an interest in control for an interest in merely recording and what is called 'explaining.'"[26] Dewey called for the development of a social psychology as a step toward the creation of techniques of social engineering. He looked forward to the realization of Condorcet's prophecy of a "future in which human arrangements would be regulated by science." For Dewey, the behavioristic movement provided the "possibility of a positive method of analyzing social phenomena." Introspective psychology (especially of the Wundtian variety), he argued, was useless for this purpose because it required that facts conform to preconceived categories. Behaviorism, he felt, "redeems us" from "such deforming of facts." In his opinion, traditional psychology had failed because it had attempted to "adapt the rubrics of introspective psychology to the facts of objective associated life." He looked forward to a time when behaviorism would "emancipate inquiry" because it represented "not an improvement in detail, but a different mode of attack."[27]

Behaviorism, Dewey believed, made possible a different world-view than that afforded by traditional concepts of psychology. As long as mind was thought of as fixed structure, he reasoned, institutions and customs could be regarded as the "inevitable result of fixed conditions of human nature." This line of argument in various guises, Dewey observed, had always been "the ultimate refuge of the stand-patter." The new view that behaviorism made possible came from its treatment of social facts, or behavior, as the subject matter of experimental science, where the problem is that of "modifying belief and desire" by initiating changes in the social environment. "The introduction of the experimental method" was, for Dewey, "all one with interest in control—in modification of the future[!]"[28]

Control of the future was a tantalizing vision that had stirred the imaginations of utopian visionaries since the beginning of the modern era. But, Dewey argued,

> . . . the need of that control at the present time is tremendously accentuated by the enormous lack of balance between existing methods of physical and social direction. The utilization of physical energies made possible by the advance of physics and chemistry has enormously complicated the industrial and political problem. The question of the distribution of economic resources, of the relationships of rich and poor was never so acute nor so portentous as it is now.[29]

It was fashionable, he noted, among those he characterized as reactionary upholders of the old order, to blame contemporary problems either on the innate wickedness of human nature or on the "bankruptcy of science" and progress. But Dewey believed that the "ultimate wickedness is lack of faith in the possibilities of intelligence applied inventively and constructively." Echoing Lippmann's theme of "drift and mastery," Dewey professed his faith in science and declared that "the recourses of a courageous humanity is to press forward . . . until we have a control of human nature comparable to our control of physical nature."[30]

Behaviorism thus arose within the context of a society whose institutions were straining under the dislocating and disintegrating forces of urbanization and industrialization. Traditional beliefs and "meta" physics were seen to be unequal to the challenge of providing the foundation of a new order. But Dewey, Lippmann, and Watson decried the pessimism of those who bemoaned the loss of the old order and called for the substitution of a new faith in the secular forces of science and technology to meet the demands of a society in which change itself was becoming institutionalized.[31]

Watson was more immediately concerned with institutionalizing change within psychology. Critics had suggested that mental images experienced while thinking, or feelings of affection, were phenomena that might have no observable behavioral manifestation and would rule out the establishment of a psychology based solely on the observation of behavior. Watson's response was dramatic and categorical. There were "no centrally initiated processes," he insisted. Instead, he proposed that the larnyx was the seat of most mental phenomena. Thought, he reasoned, was merely subvocal speech. Thinking was literally talking to one's self. He believed that the larnyx moved almost imperceptively during thought processes and that techniques could be developed that could detect and observe that activity. Affection he regarded as an "organic sensory response." Although he did not accept all of what he called Freud's "extravagances," Watson agreed with his idea of the "sex references of all behavior." On this basis, Watson suggested the notion that the phenomena of affection, or sensations of pleasantness and unpleasantness, are connected with those receptors stimulated by the tumescence and detumescence of the sex organs. Thus, he concluded, the "objective registration" of affective responses was a possibility. In this way, Watson supported what he considered to be the "essential contention of the behaviorist," that is, "the world of the physicist, the biologist, and the psychologist is the same, a world consisting of objects"—and "the method of observation of these objects is not essentially different in the three branches of science."[32]

But if the physical world of the sciences was identical, their professional worlds were vastly different. "The mistake we both made," Watson grumbled to Yerkes, "was being tied down to psychology rather than put ourselves down as zoologists." It was the lack of professional status that irked Watson. "Zoologists," he claimed, "can do the rotten type of work that [G. H.] Parker does, and Loeb, and others, and still be classed as eminent men." He felt that "with the present bunch of [university] administrators on the job . . . psychology will never come into its own."³³

Many psychologists shared Watson's frustration. Those who may have disagreed with Watson's rejection of consciousness could support his efforts to bring psychology into the scientific mainstream. In 1914, Watson received the imprimatur of his profession when he became the youngest nominee for president in the history of the American Psychological Association. He was thirty-six years old, the *wunderkind* of American psychology. He was chosen president over many of his seniors, and his election represented the coming-of-age of a new generation of American-trained psychologists, many of whom had arrived at similar conclusions about the methodology and function of psychology.³⁴ But Watson also sought and received the backing of psychologists who represented a broad range of experience and ideology. Watson won the support of the majority of American psychologists because he articulated the hopes of many in the profession who struggled for the recognition of psychology as a full-fledged member of the scientific community. He was hailed by the younger members of the profession as a "second Moses." But if Watson was looked upon as a Moses, it was not because he had delivered the Ten Commandments, but because he promised to lead the way out of the wilderness.³⁵

Despite his professional accomplishments, Watson continued to complain of pressures "that kept [his] nose to the grindstone." His wife had undergone an operation that had left his life, as he described it, "pretty badly torn up," and he protested that he would not be writing popular articles for *Harper's* if he were not "just about to perish to death!"³⁶ Nevertheless, he devoted a tremendous amount of his time and energy to his experimental work, the results of which were to have major consequences for the history of psychology. In a letter to Robert Yerkes, Watson wrote that he had been "monkeying a bit with human behaviorism." Watson had been working with his student, Karl Lashley, for almost a year on the "conditioned reflex," and he jubilantly announced that "it works so beautifully in place of introspection that . . . it deserves to be driven home; we can work on the human being as we can on animals and from the same point of view."³⁷

Yerkes, who considered himself to be "quite as much an objectivist or behaviorist" as Watson, still insisted that there was a place in psychology for introspection. Watson chided Yerkes for his reservations and maintained that for the true behaviorist, "consciousness" was no "more a scientific concept than soul." However, critics had pointed out that although Watson had advocated a behavioristic point of view, he had not offered an objective method to replace introspection. Watson hoped to answer those critics as well as overcome the reservations of sympathizers like Yerkes at the annual meeting of the American Psychological Association. "Before you decide that I am wrong about the relegating of introspection," he wrote to Yerkes, "wait until you hear my presidential address on 'the place of the Conditioned-Reflex in Psychology'!"[38]

When Watson addressed the assembled psychologists in December 1915, he confessed that he had found that "it is one thing to condemn a long-established method, but quite another thing to suggest anything in its place."[39] Yet he reported favorable results from methods tried out in his laboratory at Johns Hopkins. It was the method of the conditioned reflex developed by the Russian psychologists Pavlov and Bechterev that had drawn Watson's attention.[40] Pavlov's method of measuring conditioned responses by the salivary reflex posed technical problems for human subjects, hence Watson was inclined toward the conditioned motor reflex described by Bechterev. With this method the subject is presented with two stimuli: a sensory stimulus, such as the ringing of a bell, and a reinforcing, or "punishment" stimulus, such as an electric shock.[41] By observing and measuring the subject's response, or reflex, to the sensory stimulus with and without the reinforcing stimulus, the behaviorist could investigate such problems as "reinforcement, inhibition, fatigue, intensity of stimulation necessary to call out response under different conditions, etc." But of central importance to Watson was the fact that this method offered the possibility of an objective approach to many sensory problems that were previously thought accessible only through introspection. He noted that he gave no more instructions to his human subjects than were given to animal subjects. Nor did he care what language his subjects spoke or whether they were able to speak at all.[42] By denying that other forms of human interaction existed, Watson became detached, carefully observing the reactions of his subjects to stimuli of pleasure or pain, over which he had complete control.

Watson was exhilarated over the range of possibilities opened up by the use of conditioned reflex. Not only was it an objective method for measuring sensory response, but it had "a much wider sphere of usefulness." He had discovered that by continued conditioning it became possible to "narrow the range of stimulus to which the subject will react." The conditioned reflex then became not only a method for gathering data but also a tool by which the psychologist could modify behavior. It would become the cornerstone of a science that sought to develop techniques of behavior control.[43]

Shortly after Watson presented his findings on the conditioned reflex, he announced that he was "planning to take a pretty serious step so far as [his] own particular psychological career [was] concerned." Adolf Meyer, a leading figure in the development of American psychiatry, had established a clinic and psychological laboratory at the Johns Hopkins Medical School. He was interested in developing reliable and objective methods of studying and treating mental disorders and invited Watson to set up his laboratory at the clinic.[44] Watson had misgivings. He had earlier characterized Meyer's psychology as "pretty muddled"[45] and confessed that he and Meyer had "never hit it off very well." They had been "friendly enough, but not very helpful to each other." But Watson's apprehensions were overcome by Meyer's offer of "a magnificent suite of rooms" and the promise that Watson's work there would be "absolutely independent." While Watson did not want to neglect his animal work, he felt that his "major interests" would now "swing around to the human side."[46]

Meyer's offer appealed to Watson because he planned to spend the next several years trying to create and refine an objective methodology for deriving a behavioral theory of emotional response. Watson, characteristically, did not face the complexities of such a task with humility. "I get rather disgusted sometimes," he complained, "with trying to make the human character amenable to law."[47]

Watson worked with the confidence that he was responding to a widespread demand for methods of controlling emotions. "The psychologist is constantly being asked," he explained,

> by his own students as well as by the physician, educators, and jurists: "why do you not work upon the emotions? They are of more importance in the guidance and control of the human organism than any of your hair-splitting work upon thresholds."[48]

It appeared to some of his colleagues that Watson was "going into psychiatric work, and more or less into applied psychology in general." One contemporary speculated that the low demand for animal work in psychology had created a situation wherein those trained in that area were "forced to go into human work."[49] But Watson had always emphasized the applicability of his methods to the whole scale of animal life, and he welcomed the opportunity to begin research on higher forms. Adolf Meyer was, at first, "delighted" to have Watson "under the same roof,"[50] and Watson was equally pleased with the facilities at his disposal. He planned to install "Pavlow and Bechterew conditions" and to carry out work on the human infant. James McKeen Cattell had sent J. J. B. Morgan to work with Watson for a year, and together Watson and Morgan hoped to study the "order of appearance" and the "development of reflexes and instincts" in infants. Watson enthusiastically described his proceedings to Yerkes: "I am next door to the obstetrical ward here and I get about forty babies a month. These babies are sent over to the laboratory on demand and we can make the observations right here."[51] Watson found that the "baby campaign" was "far more difficult" than he had anticipated, but it opened up new possibilities of research for him; one of his lifelong "dreams" became the establishment of a permanent "experimental nursery."[52]

The alliance between Watson and Meyer was a fragile one, and the relationship between the two men was strained to the breaking point by a clash of viewpoints that erupted into a bitter and often vitriolic feud. As a result of his research, Watson published several articles that attempted to identify the conditioned reflex as the basis for the establishment of all emotional responses.[53] The tone of Watson's articles was brash and self-assured and, to Meyer, infuriating. Meyer's reservations about Watson's approach, and Watson's response to them, provide a glimpse into the nature of Watson's developing conceptions and a revealing insight into the personalities of both men.

Meyer's anger at Watson was mixed with disappointment and, perhaps, a sense of betrayal; for Meyer initially saw Watson as an ally in his effort to develop a theory of psychopathology that would emphasize the social and biological factors that determine personality development. Influenced by Darwin and the functional psychology of

William James, Meyer considered the mind to be the means by which an organism adapts to its environment. Meyer believed that mental illness was a result of maladaptive habits formed early in life, and he welcomed and encouraged Watson's objective study of human emotional development as a means of discovering how deviant behavior patterns are formed.[54]

Meyer had applauded Watson's critique of structural psychology because Meyer believed that abstracting "mind" into structural elements maintained a mind–body dualism that precluded any functional understanding of mental disorders. He also shared Watson's misgivings about the ability of Freudians to come up with a psychodynamic theory that could be objectively verified. But when Watson began to publish his views on the clinical applications of behavioral psychology, Meyer was outraged. Watson, Meyer thought, had merely turned structuralism on its head, substituting a mechanistic model of simple reflexes and perpetuating a dualism on the side of "body" that was no better than the structuralists' preoccupation with "mind."[55]

Meyer complained bitterly to Watson about the rigidity of his position. "You would like to see all the psycho-pathological facts treated under the paradigma of conditioned reflexes," he wrote to Watson, "with the elimination of *all* and every reference to psyche or mental, etc." Meyer could have accepted Watson's position for the sake of argument, but he felt that Watson had closed his mind to any other possibility. He complained that Watson continued "to treat those who use the word mental or psychopathological as neophytes, and [refused] any statement which deals with any other concept except motor habit." Meyer thought Watson's attitude to be "immature" and "*hopelessly* narrow." In particular, Meyer objected to what he called Watson's "exclusive pioneer tone." As Meyer told Watson:

Your application of the concept of conditioned reflexes is acceptable enough as far as it attempts to make fairly clear what the term may be made to mean; but to use it as you do as a formulation with the character of a dogma of exclusive salvation is a mere evasion of a psychophobic character, reminding me very much of the tone of the traditional "atheist" or the evolutionist a la Clevenger: it overexploits a special term from a neutral territory to make any possible reference to the old gods unnecessary, but it is not capable of any tolerance.[56]

Meyer could not abide Watson's cavalier and superior attitude. Meyer thought Watson's position to be "psychophobic" and suggested that Watson's rigidity implied something deeper than a disagreement on principle. Watson's messianic style, Meyer observed, had a shrill ring that belied the background of a man who had abandoned his rural South Carolina roots and rejected his mother's hopes that he would become a Southern Baptist preacher—only to place his faith, like many of his generation, in the salvation of a scientific materialism. In a barbed comment calculated to remind Watson that, behaviorism notwithstanding, Watson might not be so free from the past as he supposed, Meyer wrote that there

> is a point on which our upbringing differs. My forefathers have been free of the dogma of exclusive salvation since 1591; and I never had any need of eliminating a whole sphere of life interests as you did when you shed the Baptist shell. That is probably why I am much more tolerant in what I formulate as critical common sense.[57]

Meyer was particularly annoyed with Watson's use of an obfuscating terminology that masked what he considered to be a crude positivism that placed severe limits on the possibility of understanding the complexities of human experience. "Behaviorism," Meyer complained, "physiologizes the data of experience." He had more confidence in "plain language which is put to the test of daily use than . . . a jargon back of which [can be hidden] any amount of ignorance of the concepts and data of the work of other investigators."[58]

Meyer had urged Watson to modify his position before publishing the results of his experiments. He had hoped to persuade Watson by the force of reason, but he underestimated Watson's driving ambition—which precluded any compromise with competing ideas. Watson was not concerned with the data or the work of other investigators unless he found them useful for consolidating his position. As he explained to Meyer: "I have a theory of psychology which I am trying to develop into a system. I tentatively try it out first in one field and then in another. This is my sole interest." Watson already knew what he was looking for. He had outlined an all-encompassing theory that would at once explain behavior and provide the means of shaping it at will; and he was concerned with providing evidence that would support his presuppositions. Although Meyer continued to have deep misgivings about Watson's theory and method, he was sufficiently intimidated by the force of Watson's personality and professional stature to avoid an open challenge. Watson was confident of his support within the profession. His response to Meyer was calculated to demonstrate the futility of opposition.

> . . . the terminology you are using has no meaning for me; it has no meaning for [E. B.] Holt—you will notice that he has come completely over to my position and has just announced in the [Psychological] Bulletin a course in human behavior; nor do I believe that in three years time either your terminology nor psychological terminology in general will be as useful nor as helpful as you seem to think.

As to Meyer's objections, Watson smugly replied that he would "be willing to leave the matter in the hands of our colleagues."[59]

Meyer, at one point, had compared Watson's attitude with that of physiologist Jacques Loeb. "Your temperament as shown in your work," Meyer wrote to Watson, "is not unlike Loeb's. You have to shut out everything that might confuse your outlook." Meyer's comparison of Watson with Loeb is revealing; for if Watson had learned anything from Loeb while a student of his at the University of Chicago, it was that the ultimate aim of research into the nature and function of organisms should not merely be an inquiry into the nature of things but should be systematically directed toward control of the fundamental processes of life itself.[60]

Meyer was certainly not without his own presuppositions. He saw himself as part of a cultural and scientific tradition under siege by the kind of technological materialism that Watson represented. Meyer feared that Watson was wildly trampling the carefully tended vineyard that he had worked so long to preserve—hastily picking the unripe fruit to make a dangerously potent, but immature, wine. For Meyer, human nature was organically defined by both culture and biology. He viewed Watson's iconoclasm not only as scientifically suspect but also as an attack on the values and principles that he cherished.

The crux of the struggle between Watson and Meyer was that each believed the other to be leading the march of scientific progress backward into the Dark Ages. Meyer saw Watson's insistence on the physiological basis of mental disorders as a throwback to the somaticists of the nineteenth century who had believed brain lesions to be the cause of mental illness. Watson, on the other hand, considered Meyer's and the psychoanalytic movement's concern with the psychodynamics of the unconscious to be another form of metaphysics.

The thrust of Watson's research was to gain "experimental control over the whole range of emotional reactions." Watson acknowledged that much of Freud's psychoanalytic theory attempted to do the same thing, but he found Freud to be useless to the "laboratory psychologist."[61] However slight his grasp of psychoanalytic theory may have been, Watson proclaimed the "central truth" of Freud to be that

> youthful, outgrown and partially discarded habit and instinctive systems of reaction can and possibly always do influence the functioning of our adult systems of reactions and influence to a certain extent even the possibility of our forming the new habit systems which we must reasonably be expected to form.[62]

But Watson had nothing but contempt for what he called Freud's "crude vitalistic and psychological terminology." He believed that he alone had found the key to human emotional development. His 1916 experiments with infants, Watson claimed, uncovered the basic emotional reactions belonging to the "original and fundamental nature of man." These he described as fear, rage, and love. But Watson was not content merely to identify and describe these basic emotional reactions; he hoped eventually to devise the means to control their development. Fear, he discovered, could be induced by dropping the infant, by loud sounds, or by startling it when asleep. Rage was elicited by hampering the infant's movements, and love was produced by "stroking or manipulation of some erogenous zone." "Habits" or "conditioned responses" were found to be connected with these basic emotional responses at a very early stage, and these habits, he argued, not only *could* but *should* be controlled. Habits connected with the love response were especially important because "they may when not looked after sadly warp the child." Watson saw far-reaching applications for his discoveries. With the basic emotional response identified, the conditioned reflex could be used to create the desired "conditioned emotional responses." Contrasting his approach to that of psychiatrists, Watson pointed out that psychiatrists were interested largely in the patient's well-being and recovery and were not "particularly scientific." He felt that the psychologist should not be unduly concerned with the individual patient's interest when conducting experiments. Watson maintained that: "The psychologist . . . cannot afford to rest until he can control his phenomena—until he can not only *produce attachments* and study the laws of their production, but also *reduce* and break them up at will and learn the principles controlling their reduction."[63]

Watson believed that the development of such methods of regulating human conduct would prove to be of immense social utility. He longed for an opportunity to prove the practicality of his ideas, but he saw little support forthcoming from the university. An "increased financial pressure," he complained, had caused "a good deal of hard feeling and disruption" at Johns Hopkins. "There is not very much hope for us in the way of advancement now," Watson lamented to Yerkes, "or [in the] extension of the departments." But Watson, ever the entrepreneur, was quick to seize other opportunities to apply his psychological techniques. In the spring of 1916, Watson jubilantly wrote to Robert Yerkes that he had successfully persuaded the Wilmington Life Insurance Company to secure his services as a personnel consultant and was negotiating an additional contract with the Baltimore and Ohio Railroad.[64] Watson hoped that success in applying psychology to business could serve a double purpose. It could strengthen psychology within academia by demonstrating its usefulness to a large and powerful constituency and could bring the psychologist's skills to the attention of a potential market. Watson also took steps to institutionalize this partnership by offering a course in the "Psychology of Advertising" at Johns Hopkins. Designed to draw students into the university's courses in "Business Economics," Watson's course taught future managers the value of applied psychology and demonstrated to academic officials the ability of psychology to provide valuable services to the business community.[65]

Watson's enterprising interest in applied psychology was shared by many of his colleagues. In *Psychology and Industrial Efficiency* (1913), Harvard's Hugo Münsterberg presented psychology as providing the logical extension to scientific management.[66] In 1915, Walter Van Dyke Bingham had established a "Bureau of Salesmanship Research" in the Division of Applied Psychology at the Carnegie Institute of

Technology in Pittsburgh. Working with other psychologists, such as Robert M. Yerkes, Bingham hoped to develop personnel selection tests that would measure "the traits essential for success in high class salesmanship."[67] Bingham felt that "psychologists must make the results and the techniques of their science available for practical use. If we do not do this," he cautioned, "psueudo-psychologists will."[68]

Such misgivings were well-founded, since many sought to capitalize on the growing demand for the application of psychological skills. One such promoter was John J. Apatow, who, in 1916, persuaded Hugo Münsterberg, R. S. Woodwarth, E. L. Thorndike, and H. L. Hollingworth to support the establishment of an "Economic Psychology Association." Apatow confided to Bingham that he was having difficulty "making the scientific side of the organization appreciate the necessity for missionary or educational promotion work." But he had hopes that he could persuade the Associated Advertising Clubs of the World to support research similar to that of Bingham's Bureau of Salesmanship Research.[69] Bingham supported the idea of advertising research and felt that he knew of no other field "in which the application of scientific method would more surely yield large financial returns."[70] Yet he was wary of Apatow's claims to financial backing. According to its founder and promoter, the Economic Psychology Association was a result of the "interest in efficiency methods and work in scientific management." It proposed to "cover almost every phase of the human equation as it relates to business," and its goal was to "increase the productivity of the 'human material' in the average progressive business enterprise."[71] But Hugo Münsterberg also had misgivings about the financial responsibility of Apatow and threatened to withdraw his endorsement.[72] With Bingham and Münsterberg out, other psychologists backed away from Apatow's enterprise. Without their support, the Economic Psychology Association quietly collapsed. But it demonstrated an interest from both psychologists and business in such an organization and foreshadowed a later attempt at the development of a clearinghouse for such expertise by psychologists themselves. Psychologists were jealous of their hard-won reputation and were careful to make clear distinctions between themselves—who, they took pains to point out, were professionally qualified to act as psychological consultants—and those who merely claimed to have those qualifications.

Watson was alarmed enough to issue a caveat against what he called "the fake element in vocational psychology." Psychologists, he warned, had awakened too late to the problem of "the rank impostors" exploiting the field of vocational psychology. These so-called "character experts" recognized the pecuniary value of any scheme that would "enable businessmen to eliminate unfit workers and to pick out fit workers." He believed that businessmen sought traits of "aggressiveness, honesty, persistence, neatness, etc." but were unable to determine whether these qualities were present in prospective employees. Watson attributed the lack of "efficiency" in selecting employees to the tendency of managers to hire or promote on the basis of personal considerations. For Watson, "keeping and promoting the inefficient depend[ed] upon the extent to which the one who does the selecting allows personal factors to influence him." Watson recommended "performance tests" as a "scientific" solution to the problem of personnel selection. He hoped someday to offer businessmen methods to achieve the same standards of control that he had achieved in the laboratory. But he was careful to point out that only psychologists with the correct qualifications could be relied upon to provide a scientific approach to personnel management.[73]

Watson envisioned a role for psychology that went far beyond the laboratory. In 1917 he attempted to formulate a broad view of the "scope of behavior psychology."[74] This not only represented a summary of his recent work on human and applied psychology; it was also a restatement of his behaviorist position in terms of its specific social applications. Watson emphasized the application of behavioristic methodology to problems concerning human conduct. He characterized the science of psychology as having two essential functions. Through experiment and observation it attempted to formulate laws that predict how an individual or group of individuals will adjust to situations they confront in life. But beyond this, Watson felt that it was also a function of psychology "to establish laws or *principles for the control of human action* so that it can aid organized society in its endeavors to prevent failures in such adjustments."[75] Watson believed that psychologists

> should be able to guide society as to the ways in which the environment may be modified to suit the group or individual's way of acting; or when the environment cannot be modified, to show how

the individual may be molded (forced to put on new habits) to fit the environment.[76]

Although he noted that "at present" psychology had "little to do with the setting of social standards," he did insist that it could determine whether a given individual could observe those standards or devise methods that "may control him or lead him to act in harmony with them." Watson emphasized that such laws of control that may be devised must be general or comprehensive "since social standards are constantly changing."[77] Thus Watson proposed a psychology that reflected a fundamental change in the way Americans viewed their society. Small-town culture was giving way before a dynamic, urban–industrial order. Behaviorism would facilitate the adjustment of human beings to a society characterized by rapid change.

Specifically, Watson proposed a behavioral psychology that would aid businessmen in personnel selection, that would help to control crime, and that would keep men "honest and sane and their ethical and social life upon a high and well-regulated plane."[78] He pointed out that, unlike previous conceptions of psychology, behavioral psychology did not distinguish human actions from "metaphysical concepts, such as purpose, end, etc." Watson maintained that it was the function of psychology to be able to determine "whether a given individul has the reaction possibilities within him to meet the standards of that cultural age, and the most rapid way of bringing him to act in accordance with them." This task, he cautioned, would be difficult for the psychologist because he must be carefuly attuned to changing social *mores* in order to devise methods to determine individuals' adaptability to those changes.[79]

Watson's preoccupation with the practical applications of psychology were certainly not unique within the profession. But his brand of behaviorism radically departed from his predecessors and his more tradition-oriented colleagues. Whereas G. Stanley Hall, Adolf Meyer, or William James believed in a world where value, meaning, and moral order existed, Watson considered that concept to be Victorian nonsense. Social values, Watson believed, were relative to the needs of the prevailing social order. Behaviorism would make techniques of social adjustment available to those who wished to determine that order.

Watson would soon have the occasion to put his methods into practice. Amidst the rosy optimism of 1913, no one would have believed that little more than a year later the world would be engulfed in a conflagration that would shatter the Victorian order once and for all. Yet, as the United States prepared to enter the First World War, many looked upon the conflict as an opportunity to build a modern world. Here was a chance for psychologists to develop reliable techniques that would direct and control behavior. The military would also serve as a huge laboratory for the development of methods and the accumulation of data. But above all, psychologists would gain valuable experience—and make useful contacts—by cooperating with national business, industrial, and political leaders in the coordination and allocation of manpower on a massive scale.

6

Psychological Warfare

Two years ago mental engineering was the dream of a few visionaries, today it is a branch of technology which, although created by war, is evidently to be perpetuated and fostered by education and industry.

ROBERT M. YERKES[1]

There seems to have been a peculiar congeniality between the war and these men. It is as if the war and they had been waiting for each other.

RANDOLF BOURNE[2]

For many Americans, the specter of world war had loomed on the horizon ever since hostilities erupted in Europe in 1914. Woodrow Wilson had been re-elected in 1916 on a platform of peace and progressivism. Yet as German violations of America's neutrality seemed certain to compel Wilson to declare war, many of those who had supported the peace began to view the war as a vehicle for the advancement of their policies. As the United States edged closer to the brink of World War I, an influential group of American progressives seized upon the conflict as a means of transforming American society. Herbert Croly, who championed a strong central state to promote economic and political freedom, and his co-editor at *The New Republic*, Walter Lippmann, saw the war as a "rare opportunity" to advance democracy abroad and begin social reconstruction at home. By social reconstruction, Croly and Lippmann meant the substitution of rational planning for the old authorities that had been discredited or destroyed by the advent of modern industrial life. Lippmann and Croly were essentially articulating the position put forth by John Dewey, who urged that the war be used as an efficient means of achieving intelligent control over the economic and political process.[3]

Psychologists had long been in the front ranks of those who called for social engineering through applied science. John B. Watson was a leading figure in the effort to mobilize psychology for the war. He saw the approaching conflict as an opportunity for American psychologists. Watson, along with most of the psychological leadership, was not content to wait until that opportunity knocked. He suggested to Frank J. Goodnow, president of Johns Hopkins, that high government officials be contacted regarding services psychologists could provide in the event of war. Watson pointed out that an entire range of tests could be developed to select and classify military personnel. Rapid mobilization depended upon the efficient management of men and resources. Psychologists, Watson argued, had the experience and techniques to provide scientific methods of personnel selection and training.[4]

Watson was among a group of scientists and engineers who played a crucial role in developing policies that would shape relationships among science, government, and industry for years to come. As the movement toward "preparedness" gathered strength in the United States, members of the National Academy of Science, with support from the American Academy for the Advancement of Science, persuaded President Woodrow Wilson to issue an executive order creating the National Research Council. Organized in the fall of 1916, the National Research Council (NRC) was designed to coordinate research in all branches of science and to act as a clearinghouse for information. Heavily funded by private corporations, the NRC brought together governmental, industrial, and academic research organizations under the direction of a central agency.[5]

Although the immediate activities of the NRC were directed toward the war effort, its organizers looked toward the kind of postwar future envisioned by Croly, Lippmann, and Dewey. A confidential memorandum outlining the purpose of the NRC emphasized the need to stimulate the "growth of science and its application to industry." According to its founders, the NRC was organized "particularly with a view to the coordination of research agencies for the sake of enabling the United States, in spite of its democratic, individualistic organization, to bend its energies effectively toward a common purpose."[6]

The National Research Council offered those scientists and engineers who were apostles of efficiency a vehicle with which to advance

a new technocracy. Although most psychologists welcomed a chance
to demonstrate their expertise, not all could agree upon the best way
to proceed. Psychologists, led by Robert M. Yerkes (at that time
president of the American Psychological Association), argued for a
strong representation on the NRC. Others, notably Walter Dill Scott
and Walter Van Dyke Bingham (of the Bureau of Salesmanship
Research at the Carnegie Institute of Technology), wanted to provide
psychological services directly to the War Department, bypassing
involvement with the NRC. At a meeting of a special council of the
American Psychological Association called to consider the wartime
priorities of psychologists, Scott complained that the members of the
council seemed to think "of the occasion, at least in part, as an
opportunity to advance the standing of psychology." Securing a place
on the National Research Council, he admonished, "seemed in the
minds of some of the members of the council to be of more importance
than to render service to the Army."[7] Yet Scott and Bingham's self-
righteous rhetoric scarcely masked their true motivation. They were
highly interested in promoting a rating scale for personnel selection
that they had produced at the Bureau of Salesmanship Research. This
scale, initially designed for the selection of salesmen, was presented as
a method for determining candidates for officer training![8]

Many psychologists, of course, were skeptical of Scott and his
proposal. Watson, no stranger to the entrepreneurial aspects of his
profession, ironically complained of Scott's "commercializing" of
psychology.[9] James Rowland Angell worried about the "public reputation of the psychologists," of which he was "very jealous." Angell
doubted the ability of psychologists to come up with reliable personnel selection methods and measures for testing "mental efficiency"
upon such short notice. He was "extremely anxious" that psychologists not expose themselves "to ridicule or justify serious criticism."
Not the least of his concerns was the fact that any action by psychologists would come "under the severe critical scrutiny of the medical
psychiatrists." Bingham hastened to allay Angell's fears by assuring
him that his caution and skepticism was "exceeded only by that of the
members of the committee."[10]

But Angell continued to have misgivings. Although anxious to
avoid public failure, he explained that he was "equally shy of a
fictitious success which may deceive the layman for a time." He was
especially critical of the kind of publicity being given to psychology

by the press and was particularly concerned about the "utterly silly 'write-ups' of Scott's alleged discoveries" that had appeared in the midwestern and eastern newspapers. Angell worried that any short-term gains in public visibility would run the risk of going the way of all fads without the long-term support of government and industry. He urged Bingham to keep a broad view in mind as far as the profession was concerned. Referring to Scott's criticism of the oppor-tunistic attitude of Yerkes's council of the American Psychological Association, Angell maintained that "whether Yerkes was originally moved by a spirit of self-exploitation or not, matters not at all in the larger aspects of the case."[11]

What Angell, Yerkes, and others believed was that the long-term professional interests of psychologists would be better served by a secure representation on national decision-making bodies such as the National Research Council. Scott and Bingham, no less opportunistic, felt that an immediate implementation of their own techniques would win wide support for applied psychology. As it happened, after an initial period of competition, the two approaches proved to be com-plementary.[12] Indeed, many psychologists, including Watson, lent their efforts to both programs. Yerkes sought implementation of psychological testing under the auspices of the surgeon general's office, while Scott and Bingham worked to establish personnel classi-fication procedures under the War Department's auspices. The devel-opment of each program reveals much about the institutionalization of applied psychology in America.

Yerkes initially proposed that special committees be formed to consider the military application of psychology. As a result, Yerkes chaired a committee on the "psychological examination of recruits" and appointed Watson head of a committee on the "psychological problems of vocational characteristics and vocational advice."[13] Yerkes hoped that these would be incorporated into the psychology section of the NRC. This was accomplished by May of 1917, as the NRC, spurred by Wilson's declaration of war in April, began to direct its efforts toward mobilization. The psychology committee of the NRC drew its members from a cross-section of the leadership in American psychology. In addition to Yerkes, Watson, and Angell, the committee included James McKeen Cattell, Walter Dill Scott, G. Stanley Hall, Raymond Dodge, Shepherd I. Franz, Carl E. Sea-shore, Edward L. Thorndike, G. M. Whipple, and John W. Baird.[14]

After accepting the commission of major, Yerkes attempted to persuade the army to institute general intelligence tests for all military personnel. Working closely with Lewis Terman, Yerkes had developed a version of the Stanford–Binet tests that he hoped would prove useful in measuring mental ability. His efforts met stiff resistance from old-line officers, who were reluctant to see decisions concerning personnel selection pass from their hands. The military tended to view academics, at best, with caution. Respect for expertise was mixed with the suspicion that scientists considered the army to be nothing more than a vast laboratory.[15]

These fears were not unfounded, as many scientists were quite aware of the unprecedented opportunity that existed for gathering data. Some of the proposals suggested, however, indicate the ethnocentric attitudes and cultural assumptions shared by many of those who helped develop the criteria for intelligence testing—attitudes that largely influenced the interpretation of those tests. C. B. Davenport, a leader in the American eugenics movement, proposed to undertake a study of "racial differences" in military prowess. He justified his request by noting that often "this or that race" is said to be "a good fighting race" without supporting evidence. Davenport felt that a splendid opportunity existed for testing these and other assumptions. For, as he put it, there were "thousands of purebred South Italians and Russian Jews and hundreds of Greeks and Romanians, Lithuanians, etc., in the cantonments around New York" and "purebred negroes [*sic*]" in the South.[16] Such assumptions in an atmosphere of wartime crisis did nothing to mitigate the more strident campaigns for "100% Americanism." Local outbreaks of violence against certain ethnic groups that erupted during the war may not have been legitimized, but were certainly not hindered by such attitudes among the scientific elite.

Yerkes was not unsympathetic to Davenport's request and, as evidence later shows, probably shared Davenport's assumptions. But he did not want to distract the army from giving full consideration to his principal program. Yerkes assured Davenport that the surgeon general encouraged any such study of racial differences that he might "find opportunity" to make but explained that only when a "definite decision has been rendered concerning the future of our work" could a formal research proposal be considered. "Thus far," he confided, "it has seemed inexpedient to extend our inquiry for research purposes."

In order to convince the surgeon general of the value of psychology, he felt that priority should be given to those things that would immediately provide "practical and useful information."[17]

Although Yerkes did not overlook the long-term value of the data that could be collected, it was on the basis of efficiency that he promoted the army's adoption of intelligence tests. He argued that a quick and reliable procedure for measuring mental ability would permit the rapid selection of exceptional men and the elimination of those unfit for service. Two basic examinations had been developed: the "alpha" test for literates and the "beta" test for those who could not read English. Psychologists believed that they had designed a test that actually measured "native ability" rather than educational training. Yet when the results of the tests indicated a high correlation between level of education and high test scores, they took this to indicate that native intelligence is an important "conditioning factor" influencing continuance in school, instead of considering the opposite conclusion—that the tests actually reflected the influence of educational opportunity and the socialization process of the system that supported it.[18] Nevertheless, for the military, the tests had value whether or not they actually measured intelligence. The ability to quickly classify personnel according to a predetermined level of skills, however they were acquired, was of no small significance.

Yerkes found that it was one thing to win approval for his program and quite another to implement it. His effort to secure a role for psychologists within the surgeon general's office met with more than the inevitable bureaucratic inertia. The psychiatric branch of the Medical Corps saw psychologists as encroaching upon their prerogatives. With the weight of the medical profession behind them, psychiatrists within the military managed to restrict psychological examinations to the measurement of "mental ability." Psychologists were prevented from diagnosing intellectual or mental deficiency. Examination of "mental defectives" (i.e., those whose test scores put them into that category) was placed in the hands of medical officers.[19]

If Yerkes encountered administrative difficulties within the Surgeon General's Office, Walter Dill Scott and Walter Van Dyke Bingham managed to secure enthusiastic support for their program from the highest levels of the War Department. As Yerkes was devising intelligence tests and promoting their use within the Surgeon

General's Office, Scott and Bingham had gone ahead with the promotion of their own rating scale. Scott had favorably impressed Secretary of War Newton Baker and was authorized to establish the Committee on Classification of Personnel in the Army (CCPA) under the Adjutant General's Office.[20]

Although the separate approaches of Yerkes and Scott created a competition amongst psychologists for funding and government approval, both men realized that the interests of the profession demanded a united front. Consequently, the Psychology Committee of the National Research Council and the CCPA (which in fact, shared many of the same leaders) coordinated their efforts.[21]

The greatest number of psychologists who served in the government during the war did so under the auspices of the CCPA. Having secured provisional approval and an operating budget from the War Department, Scott then arranged for a scientific staff of twelve eminent psychologists to direct the Committee's various projects.[22] Many of these psychologists accepted commissions in the army. Watson, for instance, was made a major in the Signal Corps, where he, along with E. L. Thorndike, worked on methods to select and train aviators.[23]

Watson's account of his wartime experiences deals almost exclusively with his military service. The picture that he give is of a man easily frustrated by limits of any kind. Since Watson's personnel records, along with thousands of others, were destroyed by fire in the early 1970s, little remains that can verify or contradict the more sensational claims that he made almost twenty years after the war. But it is clear that once in the field, Watson found that it was one thing to devise a plan of action and quite another thing to implement it within the structure of the military. He felt "hampered," as he put it, by the restrictions of army procedure and complained bitterly about what he characterized as the "egotism" and "self-seeking" of his commanding officers. His extensive work with homing pigeons at bases in Louisiana and Texas was made obsolete by the introduction of wireless radio. He was then sent "over there" with a trunk full of questionnaires to be administered to British aviators. These were designed by E. L. Thorndike to gather data for the development of selection tests for American flight officers. Watson's destination was the Marne front, but he never made it. Heavy British losses forced cancellation of his orders. Watson was under enemy fire near Nancy.

He was shelled in Paris by German long-range artillery and witnessed Zeppelin air raids there and in London. Sailing home, he watched a torpedo cross his stern. Summing up what must have been a terrifying and confusing experience, Watson dismissed it in his characteristically cavalier fashion. "So much," he wrote, "for overseas."

Returning to Washington, Watson faced what he believed to be real danger. His true enemies turned out to be his fellow officers. Watson was nearly courtmartialed, so the story goes, because he dared to send reports critical of the "Rotation test" for aviators outside military channels. As a result of this incident, Watson dramatically insisted, he was given orders to the front to be killed in action and was saved only by the signing of the Armistice!

"The whole Army experience," Watson recalled, was "a nightmare. Never," he fumed, "have I seen such incompetence, such extravagance, such a group of overbearing, inferior men." In a backhanded slap at Scott and Bingham's officer rating scale, he criticized the selection of American officers. But his criticism had little to do with leadership. "Talk of putting a Negro in uniform!" Watson exclaimed:

> It is nothing to making a Major or Lieutenant Colonel of most of the Rotary Club men who went in as officers in the American Army (West Point and Naval Academy men excepted). The French and British officers were such a superior set of gentlemen that the contrast was pitiful; I can liken it only to a fanciful situation of a group of Yankee drummers dining at the Court of St. James.[24]

Behaviorism notwithstanding, and without Fitzgerald's irony, Watson seemed to believe, like his fictional contemporary in *The Great Gatsby*, that "a sense of the fundamental decencies is parceled out unequally at birth."

Despite Watson's bitter recollection of his military service, he emerged from the war, as did other psychologists, with an enhanced reputation. Moreover, the war broadened the horizons of professional activity. While in London, Watson came into contact with British psychologists. His work for the Signal Corps as well as his experiments conducted for the Military Intelligence Division were brought to the attention of C. E. Spearman, the director of the British Admiralty's psychological research. This was of special significance to

Walter Van Dyke Bingham, who urged the National Research Council to establish a regular liaison with British psychologists. "The British War Cabinet," Bingham wrote to Scott, "is urging the Universities once more to undertake the systematic stimulation of scientific research. Has our Cabinet a similar breadth of vision?"[25]

American psychologists hoped to stimulate such visions by focusing attention on the practical application of their wartime accomplishments. Although psychologists felt that the intelligence testing program had been less than successful from a scientific point of view, it had reaped a windfall in publicity.[26] Indeed, psychologists responsible for developing the tests encouraged a public-relations campaign to promote their program.[27] The response was overwhelming. Psychologists were besieged for information about the "alpha" test by businessmen and educators. The promise of a simple, rapid method of measuring the "mental efficiency" of large and diverse groups struck a deep and responsive chord among the nation's managers and administrators. Ironically, psychology gained recognition as a science to the degree that it removed itself from the laboratory and demonstrated its usefulness in applied fields. How to use that momentum to further the development of the profession remained an issue for psychologists to resolve.[28]

As the war drew to a close, psychologists moved to consolidate their gains. James Rowland Angell (who had helped develop the Student Army Training Corps before serving as acting president of the University of Chicago) was appointed chairman of the National Research Council. There, together with Robert Yerkes, he was able to make sure that the interests of psychologists were well represented.[29] The NRC saw its postwar role as coordinating the scientific and industrial cooperation that had been so successful during the war. Headed by Yerkes, a committee was formed to propose a long-range plan of investigation covering the area of personnel problems in industry. The issues that surfaced reflected the conservative, antiunion climate that contributed to the "Red Scare" of 1919. Among the proposals were plans to study problems of efficiency and methods of allaying worker "misapprehension" by studying the "psychology and psychiatry of trouble-makers"![30] Clearly, such terminology was intended to catch the eye of potential corporate clients rather than indicate a degree of clinical precision.

Under Yerkes's leadership, proposals were made to study the entire range of personnel problems that confronted the corporate order of the postwar world. A series of conferences held in the fall of 1919 led to the establishment of the Personnel Research Federation, which included psychologists, such as Angell and Yerkes, from the NRC; personnel experts from the American Federation of Labor, the Taylor Society, the National Bureau of Economic Research, and the U.S. Bureau of Labor Statistics; and psychologists Scott, Bingham, and Beardsley Ruml from Carnegie Institute of Technology's Bureau of Personnel Research. Thus psychologists clearly grasped the direction of the "rapidly increasing practical demands" made upon them by industrialists and administrators and seized the opportunity to assume the role of experts in problems of "mental engineering." The NRC and the Personnel Research Federation were seen as vehicles for the expansion and growth of psychology as a profession. "For the speedy and sound development of psychology as a science and as technology," wrote Yerkes, "the NRC should prove the most important of agencies."[31]

Nevertheless, those psychologists who served on the CCPA during the war believed that, in addition to strong representation on the NRC and the Personnel Research Federation, an independent organization of psychologists was needed that could market psychological services directly. The director of the CCPA, Walter Dill Scott, and committee members Walter Van Dyke Bingham and Beardsley Ruml had allied themselves with the NRC through the Personnel Research Federation, but when faced with demobilization, they sought to keep the CCPA intact under civilian auspices. Writing in January of 1919, on official War Department letterhead, committee secretary Robert C. Clothier told the members of the CCPA:

> It is our intention that the constructive work we have been doing in the Army shall not cease now that the war is over but that it shall be continued if possible in the interest of intelligent personnel direction in industry. The employment management movement has received great impetus during the last eighteen months. There is a certain responsibility devolving upon us (who have led the movement in the Army) to at least coordinate our efforts with this tendency and to aid in developing the movement in industry.[32]

Subsequently, a proposal was made to form an organization known as the Scott Company (after Walter Dill Scott). Its members were to consist of former CCPA members, including Watson, Yerkes, Angell, and E. L. Thorndike. During the war, CCPA psychologists had worked with personnel officers of many large corporations. The Scott Company soon found willing clients in such firms as the Armour Company of Chicago and the Drexel Institute in Philadelphia, where psychological tests were used in connection with a program of "foremanship research." Scott Company psychologists discovered that techniques developed during the war had a wide range of commercial applications. "It is clear," wrote Beardsley Ruml, secretary of the Scott Company, "that Army methods will be as useful in . . . the large department store as they are in the manufacturing industry." But the uses to which these methods were put were to be determined by the interests of management. Ruml characterized the Scott Company's "most constructive" work as the formulation of what he considered to be a new approach to labor relations. Ruml's "plan for an organization of workers" included a provision for a "factory research council" for the investigation of what he called "larger problems that arise in present complex industrial relationships." But these "problems," as Ruml put it, were to be considered "more from the point of view of meeting the industrial situation as it exists than trying to borrow uncritically democratic forms and formulae as they are manifested in political institutions." Ruml's message to management was unmistakable. His "new approach" to labor relations was designed to offer management a strategy to resist union organization.[33]

Another significant aspect of the Scott Company's policy was Ruml's development of the "worker-in-his-unit" concept. Ruml, who before serving on the CCPA had written his University of Chicago doctoral thesis on "Psychometry, the Study of Measuring and Determining Intelligence," proposed an approach that differed markedly from that of earlier industrial psychologists. Previously accepted theory, such as that of Hugo Münsterburg of Harvard, held that as far as men and jobs were concerned, it was largely a matter of finding square pegs for square holes, that is, to devise methods for selecting workers with the right qualifications for specific jobs. This appealed to managers who wanted to find trainable workers with appropriate aptitudes among the vast pool of unskilled, unorganized workers— thus bypassing the trade unions. This process, however, was costly and gave the trained worker potential bargaining leverage. Ruml maintained that both the worker and the job were "plastic" and should be considered together as one unit. Psychologists could develop methods that would modify both the worker *and* the job design to suit each other for greater efficiency and "unit" output. Thus the worker would become just one more interchangeable part in the production process.[34]

The efforts of psychologists to build on the foundations laid during the war began to show substantial results when the Scott Company began to attract the interest of such large concerns as the Goodyear Tire and Rubber Company, and the psychology committee of the NRC received a large grant from the Rockefeller Foundation for the development of intelligence tests.[35]

Involvement with the CCPA and the Scott Company became a springboard for professional advancement. Psychologists who had established reputations as administrators and managers found that their services were eagerly sought. In 1921, Beardsley Ruml became president of the Carnegie Corporation and, within a short time, was named director of the Laura Spelman Rockefeller Foundation. The preceding year, Walter Dill Scott was made president of Northwestern University. Other members of the company found positions as factory managers and personnel directors. Consequently, psychologists found themselves in key positions to channel influence and funds toward the development of applied psychology.[36]

The response of psychologists to the opportunties and demands of wartime had profoundly altered the discipline's status as a science and a profession. At the close of the war, Robert Yerkes could say with satisfaction that "psychology today occupies a place among the natural sciences which is newly achieved, eminently desirable, and highly gratifying to the profession." Yet its very success required that psychologists adjust to the increasing demand for applied services. "If psychology is to meet successfully the now rapidly increasing practical demands by which it is challenged," wrote Yerkes, "it must organize for cooperative endeavor in a way not thought of prior to the war. On the one hand is the imperative need of highly developed and specialized methods; on the other, the need for largely increased and adequately trained personnel." The future looked bright indeed to Yerkes. The war efforts of psychologists, he proclaimed, "have revealed or created opportunities whose scientific and practical significance cannot be estimated." Looking back at the state of the profession before the war, Yerkes acknowledged that "two years ago mental engineering was the dream of a few visionaries." "Today," he noted, "it is a branch of technology which although created by the war, is evidently to be perpetuated and fostered by education and industry."[37]

Although the war provided the arena for the emergence of a well-defined applied role for psychology, it merely served to crystallize trends that had been developing for some time. John B. Watson's earlier critiques of the underlying assumptions of psychology and his vision of a science of behavior control were instrumental in laying the groundwork for the professional structure that emerged after the war. His fight to make psychology an agent of social engineering had begun in earnest in 1913. For Watson, Herbert Croly, Walter Lippmann, John Dewey, and those who shared their hope for a world ordered by rational planning, the war became more than a campaign to make the world safe for democracy; it became a campaign against a lingering Victorian order—a war to establish a modern way of life. It was the first time in history that an entire society was mobilized for total war. From rationing to censorship from school pageants to victory gardens, it affected the daily life of the entire population. The heady taste of what could be accomplished through the creation of a technocracy was never forgotten by those who witnessed it.

There were those who were disillusioned to learn that an omnipotent government entrusted to safeguard liberty could as easily take it away. Herbert Croly, for instance, was horrified at the suppression of civil liberties during and after the war. Watson, however, despite his criticism of the army, was invigorated by his war experiences.[38] Above all, the war underscored Watson's vision of the enormous potential for applied psychology. As the war ended, Watson returned to the laboratory, where he hoped to enlarge the scope of behavior control and to develop new techniques through which that control could be applied.

7

Conditioned Emotional Reactions: Fear, Rage, and Love in Baltimore

. . . every cell I have is yours, individually and collectively.
My total reactions are positive and towards you. So likewise
each and every heart reaction.

JOHN B. WATSON to ROSALIE RAYNER[1]

John B. Watson was discharged from the army in 1918. On the surface, the America to which he returned seemed vastly different from the one he had left scarcely a year before. Although untouched by invading armies, the very landscape had changed dramatically. In a few short months, sprawling factory-cities producing vast quantities of munitions had been built in remote hollows of West Virginia and Tennessee. In Washington itself, acres of prefabricated office complexes sprouted along the Mall to house the bureaus and boards and planners who ran the war. The war mobilization revealed the astonishing degree to which a bureaucratic corporate order had already emerged in American society. Economic planning and control had been centralized in Washington beyond the wildest dreams of the most ardent progressive. Yet, ironically, the achievement of that control diminished the chances of implementing progressive social policies. Those who believed that the end of the war would usher in a new era of peace and reform were bitterly disappointed as those hopes were crushed by the Treaty of Versailles abroad and by the Red Scare at home.[2]

As historian Henry May suggested thirty years ago, the war signaled the end of American innocence about itself. But as May and others have pointed out, the war marked only the final stages of the disintegration of Victorian order. The gradual collapse of that order and the crisis of authority that it created led to ongoing attempts to establish new and lasting forms of social control. During the war, the values and ideologies of scientific management were institutionalized as part of the war production effort and the flow of information was controlled to an unprecedented degree by the Creel Committee. At the same time, the wartime crisis atmosphere loosened restraints and allowed Americans to speak and act with unaccustomed directness. The resulting clashes reflected the insecurity of the new leadership and the as yet unrefined methods for maintaining control. Calls for order inevitably contained rhetoric that was antilabor, anti-immigrant, and antipoor. When rhetoric failed, there was no hesitation to use the club. Those who benefited the most from this climate were a select few with facile, persuasive answers and the ability to implement them. John B. Watson was one of these.[3]

Soon after his discharge from the army and his return to Johns Hopkins University, Watson published a new elementary textbook, *Psychology from the Standpoint of a Behaviorist*.[4] Dedicated to James McKeen Cattell and Watson's colleague at Johns Hopkins, Adolf Meyer, the book restated many of Watson's earlier themes. But the enlarged role that Watson envisioned for psychology reflected his wartime experiences, when psychological techniques were applied on a massive scale.

Watson defined psychology as the science of "human activity *and* conduct." Moreover, Watson argued, such a science should have, like the war, a corporation, or an engineering project, clearly stated principles and goals. Psychology, Watson believed, should be able to (1) "predict human activity with reasonable certainty" and (2) formulate "laws and principles whereby man's actions can be controlled by organized society." Although such grandiose visions were not new for Watson, the war and its aftermath gave new legitimacy and urgency to his aspirations. He was quick to take advantage of the still-smoldering anti-German, antiforeign sentiment that had hounded rival industrial psychologist Hugo Münsterberg to death during the war. Behaviorism, Watson assured his readers, was "purely an American production."[5]

The war had served to highlight the achievements of applied psychology, and Watson hoped to ride the crest of that wave of enthusiasm to inaugurate his own program at Johns Hopkins. He boasted to Johns Hopkins's president, Frank J. Goodnow, that the Western Union Company had approached him to undertake the "study and standardization" of those of its employees who were considered to be "not particularly efficient." In the spring of 1920, Watson and a Baltimore physician founded the Industrial Service Corporation to market services related to personnel selection and management and to conduct "industrial psychological investigations."[6] In addition to these purely commercial ventures, Watson demonstrated the extent to which applied psychology was becoming instrumental in accomplishing the goals of a new class of social engineers. In the summer of 1919, Watson received a grant from the United States Social Hygiene Board to investigate the effects of motion pictures produced to educate the public about venereal disease. He also undertook a study on the effects of alcohol consumption during the national debate over prohibition. What were formerly moral issues regarding public obligations in the marketplace or private choices of conscience were now considered to be scientific questions to be determined by impartial observation and experimentation.[7]

Explicit in the way Watson conceived of behaviorism was the appropriation of local authority in matters of determining appropriate conduct by a growing class of specialists. The dislocation experienced by many Americans during a period of unprecedented mobility and urban growth reflected the erosion of institutions that had once provided social cohesion. Moreover, the war waged to "make the world safe for democracy" had involved a sweeping suspension of civil liberties, brought about a massive regimentation of American life, and dealt a nearly fatal blow to authority based on moral force. In 1920, a disillusioned nation repudiated Wilsonian idealism and, more than ever before, put its faith in technicians. These experts were eagerly sought out by those who were bewildered by the extent of social change and uncertain about the shape of things to come.

Watson was especially interested in creating a role for psychologists within the growing bureaucracy of education, social work, and mental health care agencies that were assuming many of the family's traditional functions. In matters of child care, for example, Watson attributed to parents a great deal of power in determining the child's emotional development. But he strongly doubted the ability of parents to use that power in constructive ways. His standards for efficiency in raising children were so far removed from reality that his "scientific" pronouncements had the effect of undermining the confidence of parents in their own abilities. This dependence upon "experts" for services formerly provided by families themselves reflected a transformation in the profile of the American family. Scarcely a generation before, most children had been raised within an extended family of aunts, uncles, and grandparents and introduced, early on, into the tasks that would define their adult lives. But in the home of the modern nuclear family, at least one parent worked outside the home. With no relatives nearby to help, the task of raising children seemed overwhelming.[8]

Watson gave scientific legitimacy to a growing sentiment among social reformers and educators that the school should take over functions of socialization formerly assumed by the family. But he went even further. Elaborating on an earlier theme, he advocated the funding of an experimental nursery that would lead to the establishment of "infant laboratories" in connection with the public school system. In this way mothers of preschool children "could be guided and warned about the way the children were tending to develop" and could receive "expert guidance and intelligent help." The laboratories would also be used to train teachers in child behavior. Watson had nothing but contempt for a "society which permits them [teachers] to *teach* instead of to *guide* the child's development."[9] Not only was Watson emphasizing the role of the school as the agent of social adjustment for the child, he was also demonstrating the importance of psychologists in facilitating that adjustment. He recommended that child psychologists take over the early grades of school instruction to insure that "many of the mishaps to the emotions due to home training could be corrected, and we could certainly be sure that from their entrance into the school system of our country no further mistakes would occur."[10]

The professional development of psychology and education had been intertwined from the beginning. G. Stanley Hall had found a natural alliance between psychology and pedagogy, and one of William James's most popular books was his *Talks to Teachers on Psychology.* Hall, James, and, later, John Dewey, were all advocates of inculcating habits, attitudes, and preparation for life—in short, conduct—as well as teaching content in the classroom. But none suggested that psychology could, or should, provide a precise formula that would allow educators to shape behavior at will. In a 1917 symposium on "Modern Science and Education," Watson argued that most biological and psychological problems center around processes of growth and development.[11] But he emphasized the need for *"predicting, controlling* and *regulating* such development."* Imparting knowledge was but incidental to the real task of public education. A partnership between the psychological laboratory and the schoolroom, Watson declared, would help develop methods that would ensure that individuals would perform according to the mandates of the prevailing social order. As Watson put it: "If it is demanded by society that a given line of conduct is desirable, the psychologist should be able with some certainty to arrange the situation or factors which will lead the individual most quickly and with the least expenditure of effort to perform that task."[12]

Ethel Sturges Dummer (a wealthy activist in the mental hygiene movement and sponsor of the symposium on education in which Watson had participated) summed up the feelings of many who looked to science and technology to provide not only solutions to social problems but also a new source of social authority. "Only through science," she wrote, "may we secure a right public opinion and better procedure."[13] Watson heartily agreed but insisted that his own definition of science prevail. The "crass reasoning" of the "Freudian mystics," he argued, "rob[s] us of our faith in our own organisms and what we may do with them." "There is no one that idolizes the organism as a whole so much as I do," he maintained, "nor who has a more hopeful conception of what we can do to that organism by training provided only society will give us suitable conditions for such training." The task of science was to discover those "suitable conditions," and, for Watson, methodological purity was of the utmost importance. The "approaches to these needed discoveries," he insisted, must be "jealously" guarded from such undesirable "shorthand methods" as the "unconscious" and "mystical" notions such as Henri Bergson's *élan vital.*[14]

Watson's self-serving disparagement of Freud and Bergson may have been intended to persuade a wealthy and impressionable benefactor, but his intransigence made him an easy target for critics of behaviorism. Adolf Meyer continued to complain of the narrowness of

Watson's viewpoint. The behaviorist, he charged in a letter to Mrs. Dummer, had eliminated "so much that we need from what he wants to see recognized as scientific psychology." In Meyer's opinion, Watson needed "a broader human outlook and balance of judgment if he is not to be as much of a danger to the development of psychology as he is a real boon and help."[15]

Watson's singlemindedness stemmed from the fact that much of his rhetoric about science was inevitably bound up with his desire for psychology to achieve independent professional goals. For too long, Watson argued, psychology had been a stepchild of philosophy. In reply to a query from Bertrand Russell, Watson made it clear that he was "trying to get psychology just as far away from philosophy as are chemistry and physics." Sharing with Russell points that he had discussed with Meyer, Watson wrote:

> One of the strongest motives I have had in trying to work a simple uncontroversial standpoint in psychology is the fact that students entering our field have to be ruined with logic clipping before they are capable of doing anything. Many of them become word artists, logicians, and pseudo-philosophers and pseudo-clinical psychologists—they will do anything which gives them a chance without being *blocked* by a system. This is the reason for the influx into the field of mental tests, trade tests and the like. But we are using up our reserve material—the world of science goes on and psychology as a science must keep not only in touch with other sciences but also work out advances in fields peculiarly its own. Hence if we are to keep our students we must have a simpler, more matter-of-fact entrance into psychology. If this is not done new practical and social applications of psychology will never be forthcoming for future use. In other words, technical or applied psychology, like applied chemistry, cannot go on long without research in the laboratory.[16]

Thus Watson made clear the connection between the development of psychology as a science and its uses as a technology. Research was inevitably linked to its application, and Watson hoped that an emphasis on practical technique would attract students and ensure the expansion of the profession.

Watson's claim that behaviorism avoided philosophical issues belied the fact that his very methodology incorporated a philosophical viewpoint. He mistook a rejection of metaphysics for a rejection of philosophy. Watson's behaviorism was a revolt against the *authority* of a philosophical point of view that had its origins in a preprofessional culture. His emphasis on methodology and results was itself a carefully reasoned philosophical position intended to replace systems no longer reflecting social realities. The attention given to technique in behaviorism was justified by Watson as a means of liberating psychology from obscure and narrow theoretical limits and making it available to a broader population (the physician, businessman, etc.). But, in reality, behaviorism only broadened the *application* of psychological techniques as instruments of control. Actual control passed into the hands of the new psychological technicians.

Despite his disclaimer, Waton's "simpler, more matter-of-fact" psychology owed much of its legitimacy to the pragmatism of William James and John Dewey. His was a psychology grounded in observed behavior and designed for use in the everyday world of action. Moreover, Watson cast his theory within a positivist framework that, if sometimes crude, appealed to many of his contemporaries, who looked to behaviorism to provide a scientific basis for their philosophical positions. Bertrand Russell's interest in behaviorism reflected Watson's growing international reputation and the appeal of his ideas. Russell asked Watson to criticize the manuscript of what was eventually published as *The Analysis of Mind* and helped to introduce behaviorism in Great Britain.[17]

Of interest to Russell and to other British philosophers and psychologists was Watson's notion of thought without consciousness. In a symposium published in the *British Journal of Psychology*, Watson discussed the question "Is thinking merely the action of language mechanisms?"[18] Although Watson had earlier claimed that thinking could be accounted for by movements of the larynx in subvocal speech, he now denied that thought was merely the action of language mechanisms. Admitting the propagandistic and evangelistic aspects of his writing, with its aim of making "converts" to behaviorism, he explained that his strong emphasis on thinking as subvocal speech was calculated to make an impression on students. Nevertheless, he reiterated his basic contention that there were no centrally initiated mental processes. For Watson, the "whole man thinks with his whole body." Thinking became for Watson a "general term to cover all subvocal behaviour."[19] Man was "only a complex of reacting systems." Problem solving was a process of trial and error but was not an isolated process. Since emotional and organic states have a great deal of influence over the process of problem solving, the "human animal" was seen as an organism that could never escape its biography.[20] Therefore, individual human beings could not be trusted to observe their own behavior accurately and, by extension, were incapable of solving their own problems objectively. Accuracy required an outside observer, preferably one who relied on instruments to record the behavior of another. Instrumentation was important to Watson because he was unsympathetic to those who tried to introduce "meaning or value" into behavior. The behaviorist must act as a detached, impartial bystander. Although he acknowledged that the very act of observing could in effect alter the observed experience, he was not concerned in the least by the dilemma posed by this inconsistency. Watson insisted that the development of experimental methodology had practical value in itself. The detached status of the observer might not necessarily lead to an increased accuracy, but it did make it easier for the observer to control the subject.[21]

This notion of control in behaviorism disturbed Bertrand Russell. Although he supported Watson in his efforts to demystify the thinking process, Russell saw potential for abuse by a technocratic elite. Exploitation of behavioristic techniques of control, he warned, could result in a society wherein an official class of "thinkers" dominated a passive class of "feelers."[22]

Russell's apprehensions were not misplaced, for Watson's research increasingly focused on discovering methods to control emotional reactions. Watson linked the changing attitudes toward the control of emotions to fundamental changes in the social structure. Since "we are no longer living in a frontier society," he argued, strong expressions of emotion formerly needed in the "struggle for existence" were no longer called for. Indeed, modern society developed "guards" against strong emotional stimuli because they interfered with the average person's efficiency. Watson felt that if emotional stimulation and emotional release could be controlled, a powerful force would be available that could be used to spur "the individual to reach a higher level of achievement." If such techniques were developed, Watson predicted, they could be used to "break through the stereotyped and habitual mode of response and arouse the individual to the point where he can accept and profit by intensive training and eliminate his errors, work longer hours, and plan his work in a more systematic manner."[23] For Watson, then, the application of emotional controls could have the effect of increasing efficiency, order, and, of most importance for an industrial society, individual productivity.

Watson's preoccupation with the control of emotions reflected his lifelong struggle with strong feelings that constantly threatened to overturn his carefully maintained equilibrium. He fervently preached, to all who would hear, the advantages of controlling one's own emotions and those of others. Most people, he argued, can cope with the demands of daily existence "if their emotions are not violently disturbed."[24] That merely to survive "the demands of daily existence" required such harnessing of the emotions is indicative of the immense power that Watson ascribed to them. Watson wanted to develop reliable methods that would channel emotional energy into predictable behavior. "Emotions," he believed, "when properly used, can be made to serve us rather than to destroy us." The terrifying specter of annihilation that he posed as the grim alternative to the subordination of emotions helps to explain his obsession with their control.[25]

Watson especially feared the unbridled release of mass emotion. One of his favorite targets for criticism was evangelical Christianity, especially as manifested in the increasing number of large, urban-centered revivals. "Every psychopathic clinic and hospital in the city feels the strain of a big revival meeting," Watson remarked to sociologist William I. Thomas. "Any type of revival meeting brings serious harm to the community," he continued. Such a volatile situation must produce "failures in adjustment" on a massive scale. These revivals must have evoked powerful memories for those city dwellers, like Watson, whose rural childhoods were permeated with strong religious overtones. Evangelical Christianity had followed the population shift to the cities, where many yearned for authentic values and beliefs in the face of the disintegrating effects of modernization. Watson, who had so ostentatiously discarded those values when he embraced modern urban life, registered his disapproval in clear, shrill tones. Religion, after all, Watson argued, was but an outmoded form of social control. In Watson's opinion, religious and political institutions were merely repositories of folkways and customs that maintained order largely through "trial and error." They were not only obstacles to rational planning but also presented clear dangers to the smooth functioning of society. The unhampered expression of strong emotion engendered by religious and political fervor was unpredictable and therefore threatening to those corporate planners who sought to maintain regular patterns of production and consumption. The modern ethic, as historian T. J. Lears suggests, called for an "instrumental rationality that desanctified the outer world of nature and the inner world of the self, reducing both to manipulable objects."[26]

Since Watson claimed to have refuted the idea of the inner world of the self once and for all, behaviorism became precisely such an instrumental rationality for manipulating the control of emotions. Watson believed that emotional reactions were due to "environmental causes" that resulted in "habit formations." If habit is the most important factor in the development of emotions, Watson reasoned, "it lies easily within our control to perfect and regulate and reshape and use practically the emotional life of the individual."[27]

This interest in determining methods by which "crude and imperfect responses" (or instincts) could be "transformed into serviceable habits" underlay Watson's turn to the study of the human infant in those years. Babies are not "hothouse plants," he argued, and can be subjected to laboratory experiments without the "slightest harm." It was in the use of infants for the study of emotions that Watson saw the greatest potential for useful results. Watson believed himself to be on the brink of discovering the basic mechanism for the development of human emotions and the key to their control. He turned to his laboratory to begin a series of experiments that were to have consequences that Watson would have been the last to predict.[28]

Before the war Watson had advanced the theory that the conditioned reflex could be used as a method of controlling emotional responses.[29] He had also experimented with human infants in an attempt to discover which emotional responses were basic and inherited and which were habits or conditioned responses.[30] The next step for Watson was to develop techniques whereby the psychologist could condition the emotions of human subjects to react in a predictable way to stimuli created and controlled by the experimenter. In the winter of 1919–1920, with his graduate student assistant Rosalie Rayner, Watson attempted such an experiment, which was destined to become the classic illustration of conditioning in the folklore of psychology.

Watson used as his subject a nine-month-old infant (Albert B.) who had been cared for in a hospital since birth. Based upon his theory that there were three basic emotional reactions (fear, rage, and love),

Watson sought to prove that these responses could be artificially induced in the subject. In calling forth the fear response by making a sudden loud noise (striking an iron bar with a hammer) and simultaneously presenting the infant with animals that had not previously elicited a negative reaction, Watson hoped to condition the subject to fear the animals even after the original stimulus (striking the iron bar) was removed.[31]

Watson claimed that the results of his experiments on "little Albert" demonstrated that emotional reactions could be conditioned to respond to stimuli arbitrarily chosen by the psychologist. Although he admitted to some qualms about conditioning an infant to fear an animal of which it was previously unafraid, Watson made no attempt to "recondition" his subject after the experiment. In fact, he speculated with some amusement that if little Albert developed a phobia to fur coats later in life, some psychiatrist would be sure to attempt to find some sexual basis for the fear. Watson argued that the conditioning of his subject through what he called the "fear" response challenged the Freudian emphasis on the primacy of sex, or "love," as the principal emotion. For Watson, the three basic emotional responses (fear, rage, and love) were equal in their potential to influence personality as well as in their ability to be conditioned.[32]

Upon close examination, Watson's experiment does not offer the kind of convincing evidence that he claimed for it. The use of only one subject, the subjectivity of the observers' accounts, and the lack of sufficient follow-up study are examples of glaring methodological flaws, not to mention ethical lapses, that critics later pointed out.[33] Yet not only were Watson's conclusions widely accepted, but the experiment itself became one of the most frequently cited in American psychology textbooks.[34] Part of the explanation lies in the appeal of Watson's behaviorism. Watson often emphasized behaviorism's potential rather than its demonstrated content, and it was behaviorism's promise that captured the imagination. The possibility of an objective science of behavior control appealed strongly to many whose progressive faith in science led them to believe that it was only a matter of time before factual and methodological problems would be solved. Yet the intensity of that faith and the suspension of disbelief in the face of such meager evidence belies a deep despair in the cohesion of the social order. William Butler Yeats was not alone among Watson's contemporaries in seeing the world as a place where "things fall apart." Behaviorism was unambiguous, straightforward, and seemed to offer a hope of certainty for those who so desperately sought it. Watson's appropriation of the conditioned reflex as a tool for the behavioral psychologist as a brilliant stroke: It made it possible for him to fulfill his original intention to extend the function of behaviorism from a psychology of prediction to a psychology of control.

By the fall of 1920, Watson's reputation was one to reckon with both at home and abroad. His professional and academic position seemed secure beyond challenge. Fearful of losing Watson to rival institutions, Johns Hopkins's president Frank J. Goodnow generously increased Watson's salary. Goodnow assured Watson that his own favorable impression of the behaviorist's work was corroborated by the faculty at Johns Hopkins. He also took pains to convey to Watson the support of the board of trustees as well as the "universal feeling" that it would be "extremely unfortunate" for Johns Hopkins should Watson decide to accept an offer from another university.[35] It was the apex of his career. Yet within a few months, Watson, to his astonishment, found himself embroiled in controversy and the subject of a scandal luridly detailed in the pages of the nation's leading daily newspapers.

Watson's crowning experimental achievement in controlling emotional responses led, ironically, to a romantic and passionate involvement with his young graduate-student assistant and co-worker, Rosalie Rayner. It was characteristic of Watson to be at once capable of sustained, concentrated, controlled behavior and impulsive, spontaneous outbursts of emotion. But, in this case, his defiance of community mores combined with the influence of the powerful Rayner family had disastrous consequences for Watson. The result was the breakup of his marriage and the end of his academic career.

Though Watson's predicament came as a shock to most of his colleagues, it came as no surprise to those who knew him well. The marriage of Watson and Mary Ickes had never been placid. Even in its early years, Watson had been involved in an affair that had left his wife distant and embittered.[36] In 1910, after he had been at Johns Hopkins for two years, Watson wrote to Angell that he had been trying to lead a "blameless life" but that his wife admitted that since the "Judson Affair" she had been "living a lie" and "no longer cared for [him]." "There is no doubt about it," Watson confessed to Angell, "she instinctively loathes my touch." At that time Watson wavered between a jealous suspicion of his wife's fidelity and a repentant assumption of the blame and guilt for their unhappiness. "Haven't we made a mess of our lives," he mused to Angell. "[For] 2 years I fought a love and conquered. Three years ago she says she loved me then the cord broke suddenly. It is all my fault." The experience left Watson feeling "weak" and "on the verge of a breakdown." He lamented to Angell that "nothing has any interest any more but I am fighting to fill my life again." Watson later reflected that it was during this time when his wife became, as he put it, "anesthetic" toward him, that he "began to accept a large part of Freud." Despite his misery, Watson assured Angell that he need not fear a "scandal." For the sake of Watson's career and the children, an "open break" was avoided. "We shall simply drift along," he wrote, "until a change comes naturally".[37]

But in 1920, after "drifting" for ten years, the break came abruptly and, for Watson, catastrophically. He had just turned forty-two when Rosalie Rayner came to study psychology at Johns Hopkins. She had graduated from Vassar the previous spring. The Rayner family occupied a prominent place in the economic and political life of Maryland. Rosalie's grandfather, William Solomon Rayner, had established the family fortune in railroads, mining, and shipbuilding, and her uncle, Isidor Rayner, had served a term in the United States Senate.[38]

It was not the first time that Watson had been unfaithful, but he saw in Rosalie's youthful adventurousness something of his own capacity for risk. "We both have the power," he wrote, "of getting what we go for." By the spring of 1920, their relationship had developed to the point where the behaviorist could declare to his protégée that "every cell I have is yours, individually and collectively. My total reactions are positive and towards you. So likewise each and every heart reaction. I can't be any more yours than I am even if a surgical operation made us one." With less clinical precision, but perhaps with more warmth, Watson asked Rosalie: "Could you kiss me for two hours right now without ever growing weary? I want you all 24 of the hours and then I'd quarrel with the universe because the days are not longer. Let's go to the North Pole where the days and nights are 6 mo. each".[39]

Watson claimed that the dissatisfaction with his marriage was shared by his wife. By Christmas of 1919, her interest in him had become "purely maternal." According to Watson, his wife complained that "married life . . . was a bore and that she hated for the evening to come." She, too, Watson suspected, had become involved in an affair and, when confronted with Watson's infidelity, declared that "Rosalie could have [him] as far as she was concerned." Watson tried, without success, to persuade his wife to take the children to live in Switzerland for a year until a divorce could be arranged, but she had other plans. She may have been willing to give Watson his freedom, but she was determined to make him pay for it.[40]

In what must have been strained circumstances, the Watsons had, for some months, become accustomed to spending several evenings a week at the Rayner home. Watson had sought to mask his affection for Rosalie behind a pretense of sociability with the Rayner family, but his wife's suspicions soon became aroused. One evening she stole into Rosalie's bedroom and spirited away a number of her husband's letters that Rosalie had concealed in a bureau drawer. These she took to her brother, John Ickes, and engaged a lawyer who persuaded her to use them as leverage in extracting concessions from her husband—and who, according to Watson, attempted to extort a considerable sum from the Rayner family. Along with testimony from witnesses who verified the liaison between her husband and Rosalie Rayner, she entered the letters as evidence in a divorce suit that did not come to trial until November, 1920. A separation agreement was reached between the two parties in late July 1920, with Mary Watson retaining custody of their two children and obtaining a substantial property and alimony settlement from her husband which left him with but a third of his former income.[41]

Watson was supremely confident that his academic reputation was secure enough to withstand any censure of his personal conduct. The matter could have passed quietly, but Watson did little to conceal his relationship with Rosalie Rayner, and it was only a matter of time before rumors began to circulate at Johns Hopkins. When Adolf Meyer learned of the situation, he urged Watson to show restraint. "I know the temper of the board of trustees," Meyer cautioned, "and would dread their attitude unless Miss Rayner's share is kept quiet." Watson paid little heed to Meyer's suggestion. Some members of the university community had known about his separation for some time, and perhaps Watson took their silence to imply a tacit acceptance of his behavior. But when the full details of the affair were revealed, Watson was asked for his resignation.[42]

The university had hoped to avoid a scandal like the one that had forced the resignation of James Mark Baldwin eleven years before. It seems that the administration was considering an alternative to dismissal. Late in September 1920, Adolf Meyer had written to Johns Hopkins's president, Frank J. Goodnow, that although it was his "genuine wish that Watson might be able to continue," he felt that "without a positive realization on the part of Watson of the total impropriety of his actions and without an unmistakable formulation of solid and binding principles for the future—the severance of relationships become inevitable."[43] The issue for the university was not only that Watson was having an extramarital affair; it was also that the affair was with a student. As Meyer put it: "Without clean cut and outspoken principles on these matters, we could not run a coeducational institution."[44] But perhaps the most important factor was that this particular student was of special concern to the university. A detail that was probably not overlooked by the administration and the board of trustees was the fact that Rosalie Rayner's grandfather had given ten thousand dollars to Johns Hopkins.[45]

If the threat of dismissal was intended to be used as leverage to force Watson to give up his affair, the attempt failed. Meyer reported to Goodnow that: "clearly . . . the conditions under which I saw some light has not been realized. There is no definite statement of principles which could justify a change of a fixed and firm attitude on the part of those who have to main[tain] the order of the university." Goodnow agreed. Although he considered Watson's plight "a very pitiful situation" that could "hardly fail to arouse our sympathy," he was not inclined to let personal considerations undermine his sense of duty. "After all," he wrote to Meyer, "we have a responsibility which we cannot evade."[46]

This was not the first time that Meyer had acted as the university's moral guardian. Meyer abhorred controversy. His characteristic response to conflict was usually an attempt at mediation. But when this failed, he tended to side with prevailing conventional opinion. Meyer feared that Watson's connection with his clinic would bring the breath of scandal uncomfortably close to his own Department of Psychiatry, which already suffered from the university's meager financial support. Meyer's squeamishness can be traced back to 1911, when (upon the heels of Baldwin's resignation) he declined to appoint Ernest Jones to a position in the department of psychiatry. Not only did Meyer oppose Jones's uncompromising support of the sexual basis of Freudian theory; he also feared that Jones's personal attitudes about sexuality would soon affront Baltimore's moral sensibilities.[47]

Goodnow and Meyer were expressing an attitude that reflected deep conflicts between public morality and personal freedom which surfaced during the progressive era. Those progressives, like Goodnow, who were concerned with the reform of social and political institutions also felt themselves to be custodians of an embattled morality. Progressive reformers considered the meaning of liberty to imply freedom from various wants and deprivations that Victorians had accepted as inevitable. But they also upheld the moral absolutes of an earlier era. Many of the social reforms of the progressive period were designed to strengthen the family, which was seen to be threatened by forces of "self-interested individualism." The family was expected to provide a socialization of the will and to enmesh the individual in a web of altruistic interests. Progress was intended to be a process that strengthened the social order and did not imply freedom from social restraint. At the time of Watson's dismissal, divorce was grudgingly beginning to be conceded as a necessary safety valve that would preserve the Protestant sexual ethic and enable the patriarchal family to accommodate itself to the pressures placed upon it by an industrialized society. Watson's divorce could have been accepted by his colleagues, albeit with difficulty, but when he persisted in his liaison with Rosalie Rayner, he overstepped bounds of permissible behavior. Even more than his conduct, it was his insistence that he had a right to persist in his actions that was particularly threatening. An ethic based on individual needs and desires posed the specter of social anarchy. Johns Hopkins might have prided itself as an institution devoted to progress, but its custodians believed that there could be no progress without moral order and "right conduct."[48]

In October 1920, Goodnow called Watson into his office for a final confrontation. Goodnow made it clear to Watson that dismissal was imminent. Watson wasted no time in bowing to the inevitable. Taking a leaf of the president's stationery, he scrawled a terse note of resignation and took his leave of Johns Hopkins.[49] It was the end of a career to which he had devoted his life. Watson was stunned. Until the end he had refused to believe that he would actually be fired. Watson was deeply embittered by his dismissal and the conditions under which he was forced to resign, but he remained steadfastly unrepentant about his personal behavior. He had been convinced that his professional stature would render him impervious to any censure of his private life. But he completely misjudged the sensibilities of the authorities at Johns Hopkins. Even so, he could not resist driving home his point in one last parting shot. Soon after his resignation, he wrote to president Goodnow:

> In establishing business connections I have been asked what I was forced to resign for. I find it rather difficult to say. I am wondering if you would be good enough to put the basis of your actions in words. I don't want it softened in the slightest degree.
>
> Nor do I wish to use it except as a plain statement of fact. I wrote certain letters to my student—there was a scandal, but when I have attempted to give the universities' [sic] basis of action I find myself at a loss and it is not due to any inability on my part to face my own actions.[50]

Although Goodnow assured Watson that "the reason for your leaving us in no way reflected upon your honesty or upon your confidence as a teacher or investigator," he warily avoided committing himself to a position. "I do not think I can very well suggest to you," he demurred to Watson, "what reasons you can give for your action in resigning."[51]

Deprived of his position and shunned by his former colleagues, Watson found himself cast adrift in a sea of mounting difficulties. It is not surprising that in the midst of this crisis he turned to his old friend William I. Thomas. Thomas, an eminent sociologist, had been dismissed from the University of Chicago under similar circumstances and was therefore in a unique position to sympathize with Watson's predicament.[52] Almost immediately after resigning, Watson left for New York, where he gratefully accepted Thomas's hospitality. There he began to piece together the strands of his unraveled life.

If Watson had hoped to secure another academic position, his hopes were dashed by the nationwide publicity that erupted when his divorce suit went to trial in November, 1920. When the press got wind of the story, it ran for a month in the leading dailies of the nation. The newspapers did not hesitate to emphasize the sensational details that were given as testimony during the divorce hearing. One of Watson's letters to Rosalie Rayner was printed in full by several papers, and reporters eagerly followed up on the story. They were especially interested in the identity of the correspondent in the case. Although the Rayner name was withheld from the proceedings, the press was quick to point out that a well-known "society girl" named "Rosalie" was the "affinity of Professor Watson."[53] But shortly after the trial, an enterprising reporter located Mrs. Watson in New York and was able to establish Rosalie Rayner's identity. Mary Watson appeared to be "good natured" and conveyed to the press the impression that she had tried as much as possible to "shield" Rosalie Rayner's name from publicity and to avoid any scandal that would endanger her husband's career. She maintained that, with the support of the Rayner family, she had tried to arrange for Rosalie to go abroad in the hope that the infatuation was temporary. When asked whether absence might have made the hearts grow fonder, she replied that, with her husband, "it would more apt to be 'out of sight, out of mind.'" The plan failed because Mrs. Watson found Rosalie "to be of a nature which seems to gain determination in the face of opposition." It was only when she found the situation to be "hopeless," she explained, that she followed the advice of her brother, Harold Ickes, and filed suit for divorce. She maintained that she bore her husband no "ill-will" and wished him well in his work. "His letters which have been made public may sound quite romantic," she said, "but [she was] quite sure they were the product of off-times in his daily routine." She added wryly that her husband did not "put romance above his work as an educator."[54]

The reactions in the academic community to Watson's dismissal were mixed. Although, after the newspaper publicity, no university could afford to risk the inevitable public outcry by hiring Watson, there was general sympathy for what one colleague called "the sad fiasco which Watson has made of his life."[55] Many agreed with W. H. Howell of the Johns Hopkins Medical School that Watson's "punishment was unnecessarily severe."[56] Perhaps E. L. Thorndike spoke for many psychologists when he said that he considered it to be his "duty" to make what he called the "quixotic" gesture of "trying to hold Watson's genius for psychology." Others, like Robert Yerkes, lost no time in letting it be known that they were available to fill the coveted position that Watson left vacant at Johns Hopkins.[57]

Watson's experience had a special meaning for those contemporaries who were concerned with the impact of changing social conditions on sexual roles and on attitudes about sexuality. Mental hygiene activist Ethel Sturges Dummer wrote to Adolf Meyer that she considered it "tragic that the 'trick of sex' should deal such blows."[58] Meyer confessed that "frankly . . . [he had] little patience with either [William I.] Thomas or Watson. Neither in their life nor in their philosophy," he wrote, "have I seen a constructive effort. I do not think that their instincts lead to anything but a mess. What on earth *did* they attain?"[59]

Although as a psychiatrist Meyer was attracted to Freud's dynamic explanation of mental disorders, he was repelled by the sexual content of Freudian theory. He had hoped to stem the tide of moral relativism, which he saw as a particularly threatening aspect of both Freudian psychoanalysis and behaviorism. When Meyer first learned of Watson's predicament, he seized the opportunity to place the blame on behaviorism's philosophical shortcomings. In a letter to Watson, Meyer intoned: "I cannot help seeing in the whole matter a practical illustration of the lack of responsibility to have a definite philosophy, the implications of not recognizing meanings, the emphasis on the emancipation of science from ethics, etc."[60]

Meyer saw Watson's behavior as an example of an emerging narcissistic style that clearly threatened the traditional family structure. "One's day-dreams, in these days," he noted, "naturally flirt at times with the alluring notions of absolute freedom." But for Meyer, sexuality, unless grounded within "the context of family formation," would lead only to chaos. Meyer supported his belief with observations he had made in his clinic. "My study of the patients and of the medical students," he observed, "makes me impatient with Watson, unless he recognizes that he has made a mess of it all. I accept him as a case of social pathology but not otherwise."[61]

Meyer was not the only one of his generation to note, with alarm, "the alluring notions of absolute freedom" that began to surface in 1920. As F. Scott Fitzgerald observed: "Only in 1920 did the veil finally fall—the Jazz Age was in flower." It was, he wrote, an era that witnessed "a whole race going hedonistic, deciding on pleasure."[62] It marked the beginning of a modern era that was characterized not so much by an enlargement of moral choice as by a shift in moral authority. The values of a nation that, for the first time, had become predominantly urban were more accurately reflected by the new consumer culture that was being manufactured on Madison

Avenue and in Hollywood than by the lingering absolutes of Victorian idealism.

For Meyer and his contemporaries, Watson's personal life and his philosophy, as Henry May has noted, came to symbolize an attempt to separate progress from traditional moral values.[63] But if Watson's adversaries condemned his conduct, even his friends began to have misgivings about his behavior. William I. Thomas, who had taken Watson in after his dismissal, soon regretted his decision. He complained that his guest was "more childish than [he] imagined."[64] Thomas wrote in exasperation that Watson "has the mother complex that the Freudians glorify, and he has it for fair."[65] He further observed that Watson's

> fault is that he expects instant appreciation and help from all who are allied with him and has no consciousness at all of reciprocity. He is like a child who expects petting and indulgence, but has no return. He would pick up a basket of chips if his mother told him to do it, but he would not pick up chips spontaneously. He thinks people have and must have a perpetual good opinion of him without regard to his behavior. . . . He has scales on his eyes, and becomes quickly a pest or a comedy to all men who know him intimately. His life is too short. On the other hand, you know his efficiency depends on these things. . . . He is a good case to watch with reference to our question whether there is any age at which habits cannot be changed.[66]

Despite his criticism, Thomas remained "very fond of Watson," although he felt that, in his professional opinions, Watson was often "a bigot."[67] Nevertheless, Thomas arranged for Watson to be interviewed by the J. Walter Thompson advertising company and introduced him to friends at the New School for Social Research. The Thompson agency had contacted Thomas earlier about the possibility of making a study of "the psychology of *appeal*." Thomas was intrigued. "I have seen enough already of their work," he wrote, "to conclude that their efforts to control the wishes of the public may have some points of value for us." He was, however, quite wary of the consequences of such efforts. "The whole advertising situation," he concluded, "seems to me bad."[68]

Watson did not share Thomas's misgivings about advertising. He was offered a job by the New School for Social Research and by the

Thompson agency but "inclined" toward the advertising position, especially after he was offered twenty-five thousand dollars for his services—a sum that more than quadrupled his academic salary.[69]

In seeking letters of reference for Watson, the J. Walter Thompson Company stressed that they were primarily concerned with his "intellectual honesty and judgment."[70] Replying to the query, Watson's colleagues and associates in academia and business readily supported his candidacy for the job. Even E. B. Titchener, Watson's staunchest critic among psychologists, had not the "slightest doubt" about Watson's "intellectual honesty and integrity." Although he felt that Watson sometimes "moved impetuously," Titchener explained this behavior by concluding that Watson "was a youth of quite unusual ability who rose in his profession with unaccustomed speed." This, he felt, may have led Watson to "act a little hastily and over-confidently" in the past, but added that Watson was now "a mature man" and "has sobered down." Even though "his temperament is aggressive," Titchener cautioned, he assured the Thompson agency that this was a quality that would be an asset in the business world. Watson was a man, he wrote, "who would never judge phlegmatically."[71]

The J. Walter Thompson Company was impressed by Watson's credentials and by the potential of applying psychological techniques to the business of selling commodities. The coming of the new year marked the beginning of a new life for Watson. After his divorce became final on December 24, 1920, he and Rosalie Rayner were married on New Year's Eve. At the same time he began a new career as an executive in the Thompson agency.[72]

It is perhaps appropriate that Watson—who believed that behaviorism would enable psychology to become useful to the educator, the physician, the jurist, and the businessman—should become a successful advertising executive. In the 1920s advertisers were instrumental in marketing the testimony of "experts" as a means of influencing public opinion. Selling new products depended, to a great extent, on the advertiser's ability to persuade the public that their habits of consumption were old-fashioned or unscientific. Ironically, it was in his postacademic career that Watson achieved his greatest influence as a psychologist. With the publication of *Behaviorism* (1924) and *The Psychological Care of Infant and Child* (1928), as well as in countless newspaper and magazine articles, Watson spread the behaviorist faith to a mass audience. He became a popularizer of psychology as a means of self-help for those who had difficulty adapting to the new social order and an advocate of psychological engineering to an emerging class of social planners and corporate managers who sought scientific methods for social control.

Adolf Meyer and the officials at Johns Hopkins were no less concerned with control, but they perceived in behaviorism and in Watson's defiance of conventional mores a threat to their own authority as arbiters of a moral order that was already under siege. Watson tirelessly campaigned for a brave new behavioristic world in which faith in science and scientific expertise would replace traditional guides for human conduct. If Meyer and Goodnow believed Watson's dismissal to be a triumph of "solid and binding principles" over "moral relativism," it was only the opening salvo in a battle that Watson was to wage over the next twenty years.

8

The Selling of a Psychologist

> . . . it can be just as thrilling to watch the growth of
> a sales curve of a new product as to watch the learning
> curve of animals or men.
>
> JOHN B. WATSON

During the bleak days when Watson's dismissal from Johns Hopkins
was imminent, he remained confident that he could find a job in the
business and commercial world. "It will not be as bad as raising
chickens or cabbages," he assured his chief critic, Adolf Meyer.
Nevertheless, Watson was loathe to abandon his laboratory and feared
that his influence on experimental psychology would be diminished.
"I feel that my work is important for psychology," Watson wrote to
Meyer, "and that the tiny flame which I have tried to keep burning
for the future of psychology will be snuffed out if I go—at least for
some time." Yet, in a characteristically categorical flourish, Watson
tried to make the best of an ignominious departure. "I shall go into
commercial work wholeheartedly," he announced, "and burn all
bridges."[1]

But Watson's bridge to the commercial world rested upon his
reputation as the founder of behaviorism. His characterization of
behaviorism as a science with broad applications to the marketplace
was particularly appealing to the J. Walter Thompson advertising
agency. Its president, Stanley Resor, was intent on transforming the
organization into a "university of advertising"—a place where scholars, scientists, and experts would set standards of efficiency and
accuracy for the industry. Resor had risen through the ranks and

bought out the J. Walter Thompson Company in 1916. In just four years, Resor and his wife, Helen, turned the somewhat moribund organization into the industry leader in total billings—a position it would keep for half a century.[2]

A graduate of Yale, Resor was one of the first college men to rise to prominence in advertising. The two intellectual influences that stayed with him were the popular lectures of William Graham Sumner and Henry Thomas Buckle's *History of Civilization in England* (1857). From these two sources, Resor constructed a world view wherein human beings were governed by irrational drives and slowly evolving folkways that were beyond reform or governmental regulation. Science, however, could, by careful observation, compile statistics that would predict mass trends. As president of the J. Walter Thompson Company, Resor put his theories into practice. He hired Paul Cherington, professor of marketing at the Harvard Business School, and made him J. Walter Thompson's director of research. A disciple of Frederick Winslow Taylor, Cherington attempted to apply the principles of scientific management to advertising by conducting endless consumer surveys to quantify buying habits. Watson, Resor hoped, could develop methods to shape those habits.[3]

"Advertising, after all," insisted Resor, "is educational work, mass education." True to his word, Resor established a training program for all new male managerial employees. Regardless of prior experience, each served a stint in every department at J. Walter Thompson, learning the functions of the various divisions before assuming the specialized duties for which he was hired. Furthermore, managers and account executives were expected to be familiar with every aspect of the production and distribution of their clients' products. Upon joining the agency, they were placed as clerks in retail shops or sent door-to-door to confront consumers with the products they were to be responsible for selling.[4] Thus it was that, late in 1920, John B. Watson found himself on a train heading south. For ten weeks he traveled through sleepy Tennessee hamlets with his samples of Yuban coffee and slogged through muddy Mississippi backroads selling U. S. Rubber boots—the epitome of those "Yankee drummers" for whom he had had so much contempt only a few years before.[5]

As he went from one small store to another, Watson found his task to be a "thankless job." He admitted that he was "shown the door quite frequently," but he found himself to be "learning at a very rapid rate even if in a different school."[6] Returning to New York, he spent another two months behind the grocery counter at Macy's before his apprenticeship was officially over. After living down the "stigma" of "being an academician," Watson wrote to Bertrand Russell, he hoped to bring his psychological training to bear on problems "connected with markets, salesmanship, public resistances, types of appeals, etc." He was in the midst of preparing what he called "a 'life program' of experimental work in advertising." On the whole, he considered himself to be "happily at work," with a wider latitude for research than he had within the university.[7] In fact, he later reflected, he "began to learn that it can be just as thrilling to watch the growth of a sales curve of a new product as to watch the learning curve of animals or men."[8]

By the spring of 1921, as Watson explained to Adolf Meyer, he was concerned with "the problems of scientific and practical control of advertising." He was highly critical of the "terrific waste" in the advertising business. "No one knows what appeals to use," he complained. "It is all a matter of 'instinctive' judgment. Whether I can establish certain principles or not remains to be seen." But advertising offered rich possibilities for applied psychology. If clients could be persuaded to provide research funds, Watson looked forward to a time when "experimentation can be carried out upon a large scale." The prospects were endless. "We can do many things," he wrote to Meyer, "which will have a very direct bearing upon human behavior."[9]

Watson joined the advertising community at a pivotal juncture in its history. The system of industrial production that had developed by the 1920s was increasingly geared toward the distribution of goods to a national market, spawning in its wake a nationally oriented advertising industry. Stanley Resor understood this perfectly. "The chief economic problem today," he observed, "is no longer the production of goods but their distribution." As the output of goods continued to grow and as population shifts made markets unpredictable, manufacturers feared economic chaos. Advertisers sought to convince businessmen that they could offer a systematic method of efficiently marketing products. But advertisers were often just as confused as their clients. As they thrashed around from one approach to another, they turned to science, and especially to psychology, to provide techniques that would rationalize the distribution and marketing process.[10]

In 1921, the National Association of Advertisers had written to applied psychologist Walter Van Dyke Bingham inquiring about the "possibility of evolving principles applicable to advertising which could be utilized as a sure guide to success in the making and placing of advertisements."[11] Watson believed that behaviorism was ideally suited for such a task. Since, for Watson, man was "nothing but an organic machine," it ought to be possible "to predict that machine's behavior and to control it as we do other machines."[12] The goal of advertising was not merely the dissemination of information about given products or services. Its purpose was the creation of a society of consumers and the control of activities of consumption. According to Watson, this could be accomplished by the use of behavioral techniques to condition emotional responses. "To get hold of your consumer," he explained to his advertising colleagues, "or better, to make your consumer react, it is only necessary to confront him with either fundamental or conditioned emotional stimuli." In order to sell a given product, one did not have to make false claims or resort to the use of "yellow copy." To insure the appropriate reaction from the consumer, Watson counseled, "tell him something that will tie [him] up with fear, something that will stir up a mild rage, that will call out an affectionate or love response, or strike at a deep psychological or habit need." These "secret and hidden springs of action" were the "powerful genii of psychology." But behavioral techniques were not to be employed randomly in the hope of hitting upon the right stimulus for the desired response. Watson had discovered that "the consumer is to the manufacturer, the department stores and the advertising agencies, what the green frog is to the physiologist." For the advertising psychologist, the marketplace was the laboratory and the consumer was the experimental animal. Consumption was a specific category of behavior, and, as such, it was an activity that could be controlled. Watson was determined to develop methods "to keep the consumer headed [his] way."[13] Using sample populations of consumers as subjects, advertisers must refine their techniques scientifically, Watson told his colleagues, "until you feel sure that when you go out on the firing line with your printed message you can aim accurately and with deadly execution."[14]

For Watson and his associates at the J. Walter Thompson Company, behaviorism seemed to offer the universal key to human motivation. A technique that would enable advertisers to influence and shape mass markets over a wide geographic distribution was considered to be essential for continued expansion. Although Watson admitted that he was troubled by the lack of individuality in the emerging mass society, his qualms did not prevent him from exploiting the situation. In fact, he said, "as an advertising man I rejoice; my bread and butter depend on it." It was the universality of human response that made behaviorism possible and its application to advertising highly desirable. "After all," Watson pointed out, "it is the emotional factor in our lives that touches off and activates our social behavior whether it is buying a cannon, a sword or a plowshare—and love, fear and rage are the same in Italy, Abyssinia and Canada."[15]

The visibility of Watson's brand of psychology within the advertising community reflected a fundamental shift in the conception and development of sales campaigns. Until around 1910, the dominant belief among advertisers was that consumers were motivated by reason or "common sense." Advertisements were designed to educate or inform the public about the usefulness of a given product. Only a small minority of advertisers argued that efforts should be made to create desires for new products. There was general agreement, however, that advertising was a combination of chance and shrewd guesses. The possibility of creating a science of advertising was considered to be improbable. Although Walter Dill Scott introduced a text on *The Theory of Advertising* in 1903, it was considered by the advertising industry to be "scientific" only in the sense that it classified the rule-of-thumb procedures that had long been in use among advertisers. Yet Scott argued that the "law of suggestion" could be employed to motivate consumers and, in 1908, his *Psychology of Advertising* discussed the importance of "emotion" and "sympathy" in influencing the consumer's suggestibility to advertising copy.[16]

After 1910, advertisers began to rely less on appeals to reason than on more indirect forms of persuasion. America was no longer a rural society with a scarcity-based economy and an ethic of self-denial. It had become urban centered. Its economy was characterized by constantly expanding production, and its values increasingly emphasized self-fulfillment. Competition within the marketplace began to shift its focus from products themselves to desires associated with those products. Not only did brands of similar products compete for markets, but advertisers vied to persuade consumers of the desirability of automobiles, for example, over competing desires for electric appliances or vacation trips. Within this context, the use of applied psychology in advertising began to grow, and the behavioristic viewpoint offered to provide further refinements in sales techniques. Behaviorism represented a departure from earlier notions of advertising psychology that had emphasized a rational appeal to what were thought to be distinct mental categories. The behavioral approach ignored questions of the rationality or irrationality of mind and emphasized instead the malleability of human behavior. In the emerging field of public relations, no less a figure than Freud's nephew, Edward Bernays, underlined this assumption. "The group mind," he wrote, "does not *think* in the strict sense of the word. In place of thoughts it has impulses, habits, emotions." Bernays urged advertisers to "make customers" such as any other commodity is produced by transforming the raw material of emotions into habits of consumption.[17]

It was indeed the promotion of style rather than substance that Watson emphasized in the marketing of products. Advertisers, he cautioned, must always keep in mind that they are selling "more than a product." There are "idea[s] to sell—prestige to sell—economy to sell—quality to sell, etc. It is never so much as dry, solid or liquid matter."[18] In the case of automobiles, for example, Watson reasoned that since all models were mechanically similar and served the same function, a constantly changing design and style that appealed to the wish fulfillment of the consumer should be the basis for sales.[19] The introduction of style into product design created an impression of novelty that rendered products unfashionable or obsolete before the end of their serviceability. In the 1920s, advertisers succeeded in creating a new symbolic universe for consumers. A universe where "real life"—the life depicted in ads—was always just out of reach. In *Babbitt* (1922), Sinclair Lewis described the world that advertising created for his hero, a small-town, midwestern businessman. "These standard advertised wares," he wrote, "—toothpastes, socks, tires, cameras, instantaneous hot-water heaters—were his symbols and proofs of excellence; at first the signs, then the substitutes, for joy and passion and wisdom."[20] By emphasizing style over substance, manufacturers hoped to create a demand for goods that were distinguished from one another by superficial differences. The research director of General Motors stated the case clearly: "The whole object of research is to keep everyone reasonably dissatisfied with what he has in order to keep the factory busy making new things."[21]

Watson employed many techniques to convey his advertising messages. One of them was the "testimonial." Testimonials had long been used by manufacturers of patent medicines but were generally held in low esteem by most advertisers. But by the 1920s, the successful use of testimonials called for a reevaluation by the industry. The J. Walter Thompson agency was the leader in the large-scale use of testimonial advertising. Under Watson's direction, the services of such notables as Queen Marie of Rumania and Mrs. Marshall Field were enlisted to endorse the cosmetic qualities of Pond's cold cream.[22] Stanley Resor defended the practice to industry critics. Testimonials were directed not to a select audience, he explained, but to the mass market. The targets of the advertiser were the tabloid readers who lived vicariously through the public personalities that were manufactured specifically for the readers' consumption. "Hero worship" he argued, was a "social fact." According to Resor, "people are eternally searching for authority." This "sense of inferiority" on the part of "the masses" was a fact that "no successful editor cares to ignore."[23]

Watson did not hesitate to use the authority of his scientific background as the basis for his own testimonials. This was accomplished through what seemed to be informative news articles or radio broadcasts that were actually intended to disseminate information relating to particular products. In one newspaper article, Watson discussed the beneficial effects of coffee as a stimulant that increased mental "efficiency."[24] In another case, a radio broadcast sponsored by Pebeco toothpaste featured Watson in a seemingly scientific discussion of salivary glands and their function in digesting food. Watson, not surprisingly, stressed the importance of brushing teeth to stimulate gland activity. But listeners who responded to an offer of additional information received a circular and samples of the sponsor's product. Watson's radio campaigns were especially effective. He was a pioneer in developing this fledgling industry as an advertising medium. His voice of authority, transmitted through a device that embodied the miracle of scientific achievement, was impressive indeed to the consumer.[25]

Watson's successful use of these techniques linked the product with an appeal to authority or a desire for emulation, but he also used what he described as "indirect testimonials." Watson's method of indirect testimonials used symbols to stimulate those responses of fear, rage, and love that he held to be the fundamental elements of all emotional reactions. Watson had found that brand appeal depended on factors other than usefulness or product reliability. Mass production rendered many competing products indistinguishable in quality and function. In one carefully controlled experiment funded by the Thompson agency, Watson found that smokers with definite brand preferences could not distinguish one brand of cigarettes from another.[26] This reinforced Watson's conviction that the marketing of goods depended not upon an appeal to reason but upon the stimulation of desire—or, as a contemporary critic of Watson's put it, "the fixation of systematized illusions in the minds of the public necessary to the use and wont of an acquisitive society."[27]

An example of Watson's use of the "indirect testimonial" was an ad campaign that he developed for Pebeco toothpaste. Watson presented the image of a seductively dressed young woman smoking a cigarette. The ad encouraged women to smoke as long as Pebeco toothpaste was used regularly. Smoking was glorified as an act of independence and assertiveness for women. Poise, attractiveness, and self-fulfillment, as well as sexuality and seduction, were linked with the consumption of cigarettes. But the advertising copy subtly raised the fear that one's attractiveness might be diminished by the effects of smoking on the breath and teeth. Toothpaste was promoted, not as contributing to health and hygiene, but as a means of heightening the sexual attraction of the user. Consumers were not merely buying toothpaste—they were buying "sex appeal." In this sense, commodities themselves became eroticized.[28]

Demographics were also an important part of Watson's strategy as he attempted to translate behavioral methodology into sales techniques. The sales platform that Watson developed for Johnson & Johnson's baby powder illustrates his approach. It also reveals much about the assumptions underlying the development of advertising campaigns. Watson intended his campaign to appeal to young middle-class mothers who were expecting their first child. The campaign was carefully designed for a demographic cross-section selected on the basis of class and race. Blacks were immediately eliminated as being "a decidedly questionable market," and in developing market information the field investigators "did not visit slum districts." Watson clearly intended to draw his market from the young, white, upwardly mobile middle class. In presenting his sales proposal, Watson made it clear that he was selling not just baby powder, but several ideas associated with the product: for example, its "purity" and "cleanliness," the dangers of infection to infants, and the desirability of using baby powder frequently. In this way Watson hoped to stimulate an anxiety, or fear, response on the part of the young mothers by creating doubts about their competence in dealing with questions of infant hygiene. Watson reinforced the implications of his message with the use of testimony by medical experts. This served the dual purpose of testifying to the "scientific" standards of the product while undermining parental self-confidence. Lacking the resources of the extended family, the isolated mother in the nuclear family increasingly depended upon the mass media for advice. Methods of child care that had passed from one generation to the next were dismissed as old-fashioned or unscientific. Advertisers were in the vanguard of mobilizing support for so-called scientific opinion in areas once dominated by folk wisdom or tradition.[29]

The values and behavior encouraged by advertising often contrasted sharply with the mores of those who had grown to maturity before the turn of the century. A society that had been production oriented and bound by an ethic of self-denial had, by the 1920s, given way to a consumption-based culture bent on self-fulfillment. The values formerly justified by tradition and community consensus no longer applied in a society wherein economic power and moral authority were increasingly dominated by bureaucratic, corporate influences. As the workplace shifted from the home and farm to the office and factory, and as schools assumed more of the socialization of children formerly provided by the family, the appearance of mass-circulation magazines, motion pictures, radio, and the automobile further eroded the authority of the family and the autonomy of the community. In their landmark study of a typical American community undertaken during the 1920s, Robert and Helen Lynd found that parents who attempted to raise children according to standards of behavior they had been raised by were accused of being guilty of the ultimate twentieth-century transgression, that is, of being old-fashioned.[30]

As a psychologist, Watson heartily agreed. The demands of a more mobile society, he wrote, made it "less . . . expedient to bring up a child in accordance with the fixed molds that our parents imposed upon us." He encouraged the development of children "who, almost from birth, [are] relatively independent of the family situation." Children must learn, wrote Watson, that in the commercial and

FIGURE 1

FIGURE 2

FIGURE 3

FIGURE 4

FIGURE 1. Emma Roe Watson named her fourth child John Broadus Watson after the Baptist theologian John Albert Broadus. (Courtesy, Mary Watson Hartley)

FIGURE 2. Pickens Butler Watson pictured here late in life. (Courtesy, James B. Watson)

FIGURE 3. John Broadus Watson, circa 1879. (Courtesy, James B. Watson)

FIGURE 4. John B. Watson's birthplace, near Greenville, South Carolina. (Courtesy, Furman University Archives)

FIGURE 5

FIGURE 6

FIGURE 7

FIGURE 5. John B. Watson at Furman University, circa 1898. (Courtesy, Furman University Archives)

FIGURE 6. As an instructor at the University of Chicago, Watson met Mary Ickes, whom he wed. In June 1905 their first child Mary was born. (Courtesy, Cedric Larson)

FIGURE 7. Mary Ickes Watson is shown outside their Baltimore home with John, at age 2, and Mary, at about age 4. (Courtesy, Mary Watson Hartley)

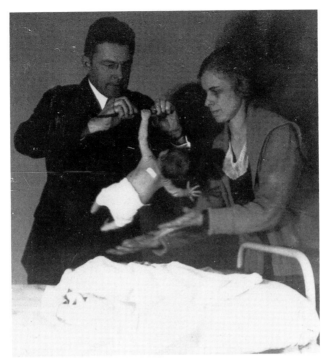

FIGURE 8

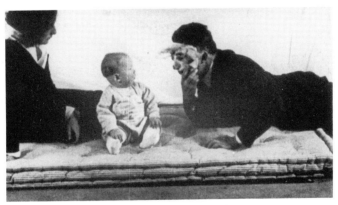

FIGURE 9

FIGURE 8. Watson began experiments on human infants as early as 1916. Here he tests the tonic grasp reflex in newborns with an unknown assistant. (Courtesy, Johns Hopkins University)

FIGURE 9. This photo shows Watson, at right, during his famous "Little Albert" experiment. At left is Rosalie Rayner. (Courtesy, Ben Harris)

FIGURE 10

FIGURE 11

FIGURE 12

FIGURE 13

FIGURE 10. This picture was taken in 1920 when Watson was 42 years old. (Courtesy, Mary Watson Hartley)

FIGURE 11. Mary Ickes Watson around the time of her divorce from her husband. She was 37 years old. (Courtesy, Mary Watson Hartley)

FIGURE 12. Rosalie Rayner was a student at Vassar when this picture was taken around 1919. (Courtesy, Cedric Larson)

FIGURE 13. Some of Watson's love letters to Rosalie were obtained by his wife who used them as evidence in a highly publicized divorce suit. (Circuit Court of Baltimore)

FIGURE 14

FIGURE 15   FIGURE 16

FIGURE 14. *The New Yorker* called Watson the "chief showpiece" of the Thompson Company. He is shown here in 1930 with Clarence Darrow. (Courtesy, J. Walter Thompson Company Archives)

FIGURE 15. Touted as "America's greatest child psychologist," Watson published his best-selling book on child care in 1928.

FIGURE 16. In 1935, Watson joined the William Esty Agency where he remained as vice-president until his retirement in 1945. (Courtesy, Cedric Larson)

FIGURE 17

FIGURE 18

FIGURE 19

FIGURE 20

FIGURE 17. During the 1930s, Watson commuted to Manhattan every day from his estate in southern Connecticut. (Courtesy, Mary Watson Hartley)

FIGURE 18. Watson built a working farm on his Connecticut estate, complete with livestock, a barn, and piggery. (Courtesy, Cedric Larson)

FIGURE 19. Watson, ever the dandy, also delighted in regaling his guests with hard-drinking, backwoods manners. (Courtesy, James B. Watson)

FIGURE 20. Watson refused to drive an automobile, but raced his speedboat, "Utopia," on Long Island Sound. (Courtesy, James B. Watson)

FIGURE 21   FIGURE 22

FIGURE 23

FIGURE 21. John B. and Rosalie Rayner Watson at the Longshore Beach Club in Westport, Connecticut, 1930. (Courtesy, James B. Watson)
FIGURE 22. Watson's two sons by his second marriage: James, at left, and William, the eldest, at right. (Courtesy, James B. Watson)
FIGURE 23. After his retirement, Watson became more and more of a recluse. His favorite companions were the animals on his estate. He is shown here with his horse "Shadow." (Courtesy, James B. Watson)

FIGURE 24

FIGURE 25

FIGURE 24. Watson sold his estate in the early 1950s and spent the last years of his life on a small farm in western Connecticut, strikingly reminiscent of his boyhood home. (Courtesy, Mary Watson Hartley)

FIGURE 25. Watson busied himself with the ritualistic task of caring for animals and maintained a workshop in the barn where he slept during the summers. He is shown here at age 80, shortly before his death in 1958. (Courtesy, Cedric Larson)

industrial world, "there is no one there to baby us."[31] The modern child would soon learn that real authority lay not in the family but in the marketplace and in its supporting social institutions. Achieving success depended upon internalizing the values of the corporate order. Success itself came more and more to be seen as the ability to emulate a style of living defined and exemplified by mass advertising.

In a world where the image was all-important, Watson reigned as the "chief show piece" of the Thompson agency. Stanley Resor had installed its offices in the new Graybar Building, adjacent to Grand Central Station. Watson and his colleagues could commute from the suburbs, lunch in the executive dining room furnished with Colonial antiques, and return home in the club car after a day's work—all without once stepping onto the sidewalks of New York. Beyond a wrought-iron cage of Spanish grillwork, Watson's lavish executive suite afforded a view of crowds of people far below as they made their way along Lexington Avenue toward the maze of streets in downtown Manhattan. From those heights, the prospects of applied psychology must have appeared bright indeed.[32] "I believe you will be happier in business," he wrote to Robert Yerkes. "I did not think that I would be," he confided "but now I would not go back for the world." Yet Watson maintained contact with his former academic colleagues. He consulted with them on advertising research projects and looked to them to provide young Ph.Ds as recruits in the growing army of market researchers.[33]

Watson did not confine his business interests to advertising, nor was he alone among psychologists in envisioning a tremendous commercial potential for applied psychology. In 1921, at the invitation of James McKeen Cattell, Watson joined the most respected names in American psychology on the board of directors of the Psychological Corporation. The Psychological Corporation had been formed soon after World War I to follow up on the successes of the Scott Company. It was to function as a clearinghouse for marketing psychological expertise. Robert Yerkes put it bluntly: "It is urgently important," he wrote to Cattell, "to place the control of psycho-technology in the hands of professional psychologists and to insure its staying there." After a period of trying to market a revamped version of the army "alpha" test as a basis for industrial personnel selection, the Psychological Corporation found its market surveys—particularly the "Psychological Sales Barometer"—to be extremely successful.[34]

In 1932, Henry C. Link (later secretary of the Psychological Corporation) published *The New Psychology of Selling and Advertising*, which included an introduction by Watson. Link noted that a "revolution" had occurred in the method of distributing goods. The consolidation of manufacturing industries and retail outlets indicated by the growth of chain stores had had the effect of eliminating the middleman (the small, independent retail outlet) and had opened up vast markets for the distribution of consumer goods. Advertisers had previously been concerned with overcoming "sales resistance," but Link saw the emergence of a sophisticated system of market research, production, and distribution that would minimize such unpredictable variables. The key element in this process, wrote Link, was psychology. Psychologists could help manufacturers "discover and sell articles to which there will be the least resistance" and help advertisers "crystallize the latest wants of consumers into active demand." In his introduction to Link's book, Watson described the extent to which psychology had already become established in advertising. Psychology had moved out of the laboratory and into the marketplace. Market research had become an integral part of advertising campaigns, and advertisers had begun to institute their own laboratory studies to test consumer reactions. The "science" of advertising, said Watson, was the "psychology of selling," and advertising had become scientific in the extent to which it had adopted psychological methods.[35]

Watson also argued that behavioral psychology could be useful as a tool in personnel management. He contributed to the training of salesmen in his own and in his clients' companies. Watson maintained that tests could be devised that would measure the performance of office workers. These tests would serve two purposes. By plotting a curve of office production, employees' output could be measured and controlled. A method of measuring the productivity of individual employees would also retard the emergence of any group solidarity among office workers that might limit production and management control. Watson was arguing for a role for psychologists within the management structure itself. Heretofore psychologists had acted chiefly as consultants in the preparation of personnel selection tests. For Watson, these tests were useful only as a rough screening device. Although psychological tests "may help us to separate the sheep from the goats," he wrote, "they will not tell us much about the flock of sheep left from which to make our individual selections." In his estimation, "men and women fail in their jobs, not from lack of intelligence . . . but because of faulty emotional organization." Watson's solution to this problem was to deemphasize personnel selection and to stress, instead, control over the employee—whether on the assembly line or behind a desk. As a behaviorist, Watson believed that management techniques could be devised that would produce efficient, well-controlled labor from a random selection of workers. A psychologist on the staff of a business organization could standardize the production of office work and assign it to employees in units so that work efficiency could be measured. In words reminiscent of Frederick W. Taylor, Watson argued for the extension of scientific management from the shop floor to the office. "[T]he main problem to be solved in the office," Watson wrote, "is the problem which has already been solved in the factory, or is in the process of being solved there. It is the problem of getting units of work comparable with the piece work of the factory."[36]

But control over work flow and office routine was not enough. Watson insisted that successful management depended upon the ability to motivate broad patterns of behavior. Traits considered essential for the successful bureaucrat were neatness, loyalty, subordination, a passive temperament, and compatibility with co-workers. But above all, the employee's job should mean more to him than a service rendered in return for wages. It should represent "a real integral, vital part of . . . life."[37] In short, the psychologist's task was to enable management to create "the organizational man" who identified his wishes with those of the corporation and subordinated his life goals to its demands.

Watson emphasized that, beyond any other considerations, the "first duty" of the businessman is "to sell." In order to be in command of a selling situation, Watson advised, a businessman should be in control of his own behavior and be able to predict the behavior of others. In an increasingly complex society mere friendliness and sincerity were not enough to be effective. "Today," Watson declared, "we have got to sell ourselves to other people just as we would sell a commodity." In terms that any Main Street booster would understand, Watson argued that since human beings were biological machines, a businessman should be able to take an inventory of himself as he would do for any other machine and be able to sell those qualities the same way he would sell an automobile. A knowledge of the science of human behavior would also give salesmen a manipulative edge over clients and business opponents.[38]

Behavioristic rhetoric aside, Watson was peddling the new ethic of "impression management"—the art of manipulating others by manipulating the self. The creation of the "other-directed man" was celebrated by Dale Carnegie and Watson's advertising colleague, Bruce Barton, among countless others. But as success became dependent upon the creation of an evanescent image, self-control, formerly an act of moral choice, became merely another tool for secular achievement. With no clear core of self, selfhood itself began to lose coherence. Fact and fiction mingled freely. "You resemble the advertisement of the man," Daisy Buchanan remarked to Jay Gatsby in Fitzgerald's novel of the Jazz Age. Advertisers themselves were not exempt from such a fate. As Watson succeeded in creating talismans of success with which to sell his products and ideas, he could not escape being evaluated in their image.[39]

In 1924 Watson was made a vice-president in the Thompson agency. "You place a sort of economic sanction on behaviorism," wrote E. G. Boring of Harvard, "by rising to business heights which must transcend the attainment of any other psychologist."[40] Watson's success reflected the establishment of a consumer culture as the dominant pattern of twentieth-century life. He helped to legitimize a new hierarchy of authority by developing advertising campaigns that associated products with self-fulfillment and by promoting the role of scientific experts in the production and marketing process and as guides to consumption. As social goals receded beyond the political horizon, scientific experts helped sanction the pursuit of efficiency and gratification as ends in themselves. Watson's old friend, sociologist William I. Thomas, shrewdly observed that scientists who become foundation directors or who serve on boards of directors:

> follow the plan of promoting studies that are "scientific" and not "social." [This] means that those who have great financial interests at stake wish studies which promote change and progress in material things but wish social conditions to remain undisturbed.[41]

As an advertising psychologist, Watson had come to recognize "the importance of maxims—how potently cut and dried verbal formulae serve as stimuli for shaping our own reactions. This is especially true," he wrote, "when those formulae are handed down by persons in authority."[42] Advertising, Watson explained, had opened his eyes "to how simply things can be put to the public, how nearly all the worthwhile truths of science can be set forth in homely words."[43] Watson had found his calling. As he became recognized as a popularizer of psychology, he used all the advertising techniques at his disposal to sell behaviorism to a mass audience.

9

Designing Behavior

*I am trying to dangle a stimulus in front
of you, a verbal stimulus which if acted
upon, will gradually change this universe.*
JOHN B. WATSON[1]

In 1926, William I. Thomas remarked that "Watson has nothing new. He is living on his past and writing it up for the magazines."[2] Watson's success at publishing for the mass market earned him a certain amount of notoriety among his former colleagues and associates. But it was, ironically, through his skill as a publicist that behaviorism made its strongest impact. During the 1920s and 1930s, Watson not only brought behaviorism to the attention of a broad general readership but also sparked lively debate among wider scientific circles. Through an enormous output of books, magazine articles, newspaper stories, and radio broadcasts, he was able to establish himself in the public eye as an expert on subjects ranging from child rearing to economics. Preaching a gospel of achievement through self-control, Watson became the first "pop" psychologist to the newly urbanized middle classes. His vision of a behavioristic future articulated a common belief in the blessings of a technocratic society and a longing for order and efficiency.

But Watson's turn to popular writing was hardly a recent development, nor was it a departure from professional standards. He had written for mass-circulation magazines since his days at the University of Chicago, and it was a tradition distinguished by the likes of William James and G. Stanley Hall. Watson had found, like James and

Hall, that psychology needed a public-relations dimension if it was to be acknowledged as a full-fledged science. But even when, after his dismissal from Johns Hopkins, he began to capitalize upon his skills as a popularizer, he never completely abandoned hope of reestablishing his research. He struggled, in fact, to find a way to continue his experimental work and maintain his academic connections. During the 1920s he taught at the New School for Social Research and supervised experiments on infant behavior at Columbia. He also continued to refine his theory of behaviorism.

In 1922, Watson reviewed Bertrand Russell's *Analysis of Mind* for *The Dial*.[3] Russell had applauded Watson's attempt to move psychology out of the realm of introspection into a study of observed behavior. But Russell maintained that there were mental phenomena that could not be accounted for by a strict behaviorism that totally denied the existence of consciousness. Watson sent Russell a page proof of his review and confessed to Russell that "there is no question but that you have forced me to make an admission about the 'image' which . . . I had known for some time that I would have to make."[4] Russell had argued that images of an object that persist after an object is perceived or when the perceived object is no longer present contradicted Watson's claim that there were no centrally aroused mental processes. Watson admitted to Russell that he had deliberately concealed these inconsistencies in the initial presentation of his theory of behaviorism; the force of his argument was intended to be more rhetorical than scientific. He had wanted, he explained, to avoid dealing with problems like the "image" "until [he] had forced psychologists away from their old point of view."[5]

Watson conceded that he had "yielded a good bit of ground but with a string to it."[6] In his review, he admitted the existence of images but held that they could be explained in purely physiological terms. Images were merely the result of excessive stimuli, Watson reasoned, such as tension on the retina caused by minute adjustments of the eye muscle. For Watson there was "no purely mental world, only the world of 'sensation' which is common property both to physics and psychology—the *neutral stuff* out of which both are constructed."[7]

The orientation of Watson's methodology attracted positivist apologists like Russell who took Watson's rejection of introspection as essentially a rejection of "unobservables." "Behaviorism," Russell wrote in 1923, "is in the first instance a method in psychology, and only derivatively a psychological theory." Although "it is possible to accept the method without accepting the theory," he pointed out, "the one leads by a natural development to the other." As a method, Russell explained, behaviorism is characterized by a rejection of introspection as a special source of knowledge about "mental" process. Behaviorism appealed to Russell because it eliminated the mind-body dualism that had preoccupied philosophical speculation for centuries. It held that the sensory data responsible for our perceptions of the external world did not differ in kind from those that were responsible for emotional states. There existed only one form of perception and its source was sensory stimulation, which could be observed, measured, and controlled. There were no independent "mental states" that did not have their origins in external sensory stimuli.[8] For Watson, however, states of consciousness were not merely unverifiable; they did not exist. This categorical assertion required a leap of faith that few psychologists were willing to make. But Watson's vision of a science that could predict and control human behavior fired the imaginations of an emerging group of professionals who began to see themselves as "behavioral scientists."

In 1924, E. B. Titchener, one of Watson's more vocal critics, described the state of American psychology to a Russian colleague in the Soviet Union. "Behaviorism," Titchener observed, "has spread over the country in a great wave."[9] Watson may have been carried on the crest of that wave throughout the 1920s, but, among some circles, his extreme brand of behaviorism was considered to be merely the surface of a much deeper current. "We must," Robert M. Yerkes cautioned William I. Thomas, "be careful to avoid confusion of the Watsonian cult with the scientific study of vital phenomena, including both behavior and experience." Thomas assured Yerkes that his use of the term *behavioristic* was not meant "in the Watsonian sense." It was "too bad," he lamented, "that a perfectly good word should be appropriated by this school."[10]

Yerkes and Thomas were commenting upon the broad movement, which gained momentum during World War I, to organize the natural sciences around practical sciences of human behavior. Despite enormous advances in industrial efficiency, social crises involving labor unrest, radicalism, crime, delinquency, and so forth seemed resistant to solutions. It was the intractability of the "human factor" that the professional managerial classes saw as the last obstacle to a rationally managed society. Within this context, human behavior became the focus of a new synthesis in an increasingly integrated research community. As anthropologist Clark Wissler noted in 1925, "there is setting in a drift toward the human behavior problem that will bring a new line up in the social, psychological and anthropological sciences." This new configuration, he remarked to James Rowland Angell, will be characterized by "a true community centering around the human behavior problem." No one knew better than Angell and Yerkes that this reorientation of scientific research was hardly due to a "drift." Through their leadership in such organizations as the National Research Council, the Carnegie Institution, and the Psychological Corporation, to name a few, Yerkes and Angell worked hard to ensure precisely the result that Wissler had observed. In the years following World War I, a complex network of interrelationships among research councils, foundations, government bureaus, trade and business associations, and universities developed that fostered the growth of behavior-oriented research.[11]

Watson's claims to have already identified the mechanisms underlying human behavior and the techniques to control it undercut Yerkes's efforts to establish a broad base of support for continued long-range research—hence his care in distinguishing the work he supported from Watson's rhetorical extravagances. At the same time, however, behaviorism's visibility helped secure support for some programs and lent a measure of legitimacy to others. Watson's work at Columbia for the Laura Spelman Rockefeller Memorial Fund is a case in point.

One of the most important factors influencing the direction of scientific research in the 1920s was the transformation of large foundations from charitable trusts to corporate, professionally managed enterprises. The Rockefeller Foundation—in particular, the Laura Spelman Rockefeller Memorial Fund (LSRM)—played a large part in shaping the direction of behavioral research. The Director of the LSRM was Beardsley Ruml. Ruml, a protégé of James Rowland Angell, had obtained his doctorate in "psychometrics" (mental testing) at the University of Chicago and had served with Watson and Yerkes on the Committee on Classification of Personnel in the Army. As that committee evolved into the Scott Company (forerunner of the Psychological Corporation) following the war, Ruml became its executive secretary and, as such, developed and marketed programs in applied psychology.[12] Ruml carried his support of applied science over into his leadership of the Laura Spelman Rockefeller Memorial Fund. The program that he developed, Ruml explained, was dominated by a "practical motive." He wanted to support research that would, as he put it, "achieve concrete improvement in the conditions of life" Ruml made it clear that "the Memorial had no interest in the promotion of scientific research as an end in itself." As foundation support became more "problem" rather than "discipline" oriented, scientific research came to be defined more and more by issues concerning social conduct and behavior.[13]

It was small wonder, then, that the Laura Spelman Rockefeller Memorial Fund picked Watson to supervise a series of experiments on preschool children for the Manhattan Day Nursery at Teachers College of Columbia University. In outlining his research, Watson insisted that "if the infant work is to be of any distinct value . . . its results must be capable of application in the home." By this Watson meant the development of procedures by which parents and teachers could control the behavior of children and shape the characteristics of their personalities. The so-called problems that Watson proposed to study, however, carried a distinct Victorian cast despite all the behavioristic jargon. "Thumb sucking," Watson warned, bred "introversion, dependent individuals, and possibly confirmed masturbators." The "uncontrollable child" (from ages two to four) was a result of "bad handling" through a series of "negative conditioned reflexes," which Watson believed could be removed. Watson also suggested experiments on the conditioning of fears in children, tests on "the whole range of stimuli calling out emotional reactions," and studies of incontinence and masturbation. As if to justify his argument or substantiate his line of reasoning, Watson threw in a bit of Freudian terminology, however loosely used or dimly perceived. As part of his proposal, he suggested the study of what he called "mother or father 'transference'" and "the best methods for breaking it up." The lives of parents, Watson held, "are made a burden by these transfers—they cannot get out of the sight of the child . . . without creating a disturbance. Handling transfers the right way at an early age may save the youngster a lifetime of misery." Such a narrow concept of parenthood, not to mention childhood, may partly explain the draconian measures Watson recommended to devise methods for teaching children "to let objects alone" without the constant admonition of parents. "Would it not be possible," he suggested, "to arrange a table containing interesting but not to be touched objects with electric wires so that an electrical shock is given when the table to be avoided is touched?" Completely ignoring the consequences for the child, Watson explained that the point of this experiment was to devise methods of controlling behavior "without having the parents as the main conditioning factor."[14]

Rather than recoiling from such suggestions, the Laura Spelman Rockefeller Memorial Fund viewed Watson's proposal as an unsentimental, therefore scientific, way of studying human conduct—the first step in the engineering of social relationships. In effect, Watson was attempting to create techniques that would reduce child rearing to standardized formulae. As the nuclear family became the norm, the burden of raising children increasingly fell on households in which the father's job demanded that he spend most of his time outside the home and the mother's time was divided between attention to "labor-saving" devices and the supervision of children. Like the school, the modern home was designed to operate along lines of efficiency modeled by industrial production. The creation of methods that would produce docile, obedient children with a minimum of attention and parental decision making was considered to be a legitimate object of research.

In 1924 Watson wrote to his former colleague, H. S. Jennings, that "my academic interests are still very strong—too strong to be given up; so I am effecting a very happy compromise by working all day in business and most of the nights in academic work." The "Rockefeller interests," Watson reported to Jennings, had given Teachers College of Columbia University a grant of $15,000 to continue the infant work he had begun at Johns Hopkins.[15]

Watson suggested that Mary Cover Jones, a graduate student in psychology at Columbia and a friend of his young wife, Rosalie, undertake the actual research and that he act as her supervisor. The experiments were a continuation of his research at Johns Hopkins on conditioned fear responses in children. Whereas Watson had previously claimed to have induced specific fears in children by behavioral methods, these new experiments attempted to demonstrate that fears in children could be reconditioned or, as Watson put it, "unconditioned" by applying the same principles. Although he cautioned that his conclusions were based upon preliminary evidence, he believed that "the whole field of emotions, when thus experimentally approached . . . opens up real vistas of practical application in the home and in the school—even in everyday life."[16]

Watson found a welcome reception for his ideas at The New School for Social Research. Founded in 1917 by Charles A. Beard and James Harvey Robinson, The New School ws intended to embody the progressive ideals articulated by Walter Lippmann, Herbert Croly, Thorstein Veblen, and John Dewey. It was conceived as a center of research and planning that would provide the blueprints for reconstructing society. Its faculty would design the mechanisms for engineering social change. Theory and practice would be united. Students would come from all walks of life to become engineers, architects, and social workers—the agents of progressive reform. In reality, however, the lofty hopes of underwriting research projects had to give way to the more practical matter of providing a solid base of financial support. In 1922, Alvin Johnson became the director of The New School and, as such, initiated a program of adult education that became a mainstay of funding for the institution and a key feature of its program. Johnson immediately recruited a stable of popular lecturers likely to draw tuition-paying students. Along with Louis Mumford and Harry Elmer Barnes, Watson was among Johnson's first choices. Watson gladly accepted the lectureship. For him it was to be "a gradual return to respectability."[17]

During the 1920s, the New School was housed in a row of pleasantly shabby brownstones in the Chelsea section of Manhattan. The makeshift arrangement of offices and classrooms made for a sometimes boisterous but convivial atmosphere for teachers and students. Watson was, perhaps, relieved to find that despite the *avant-garde* reputation of The New School, his classes were not filled with radicals and bohemians. A self-styled iconoclast on issues of his own choosing, Watson was often intolerant of other dissenters. He characterized his students as "intelligent people" consisting of "social workers and the like, . . . an ordinary bunch." Above all, he wrote, "they are not Bolshevists."[18] Watson's lectures were enormously popular. Alvin Johnson, of course, was delighted. "I doubt very much," he wrote, "that you can find anywhere in the country a comparable body of mature persons seriously interested in psychological science, especially in psychology in its bearings on behavior."[19]

Other members of The New School's faculty were not as enthusiastic. After listening to one of Watson's lectures, Thorstein Veblen grumbled that Watson would "never know as much as Dewey and James forgot."[20]

Watson gave a series of weekly lectures at The New School from 1922 until 1926. In describing his course, Watson explained that "behaviorism states frankly that its goal is the gathering of facts necessary to enable it to predict and to control human behavior." He also pointed out that "the implications of behaviorism are definitely social and that its findings must be taken into account by sociologists, economists, historians, physicians, and political theorists."[21] To emphasize this point, Watson used his forum at The New School as a platform upon which to engage his colleagues in an extended debate on theoretical and professional issues. In 1924 Watson offered a course in "Human and Infra-Human Behavior, A Study of the Evolution of Habit Formation," to which he invited leading geneticists and biologists (T. H. Morgan, C. B. Davenport, and H. S. Jennings) to speak.[22] Watson had remarked to Jennings that he had been spending "a lot of time upon instincts. The concept of instincts in man I think is weakening considerably. The work I have done on man, including the experimental work I am now doing at the Heckscher Foundation, makes me feel that inherited patterns in man are almost at a minimum. Almost everything seems to me to be built in."[23]

The instinct theory of human behavior had long influenced American psychologists, biologists, and social scientists. As formulated and articulated by George Romanes, William James, and William McDougall, consciousness was said to exist in all animate organisms—the complexity of that consciousness depending on the complexity of each species's neurological structure. At each stage of evolution consciousness became more complex, resulting in an unbroken chain of mental development from the lowest organism to man. Certain behavior patterns in man were seen as vestiges of behavior observed in animals and were characterized as instincts transmitted from one generation to another by heredity. This theory, though unproven, was largely accepted as fact until the results of experiments began to challenge those assumptions.[24]

William McDougall, a British physician, applied the instinct theory to human social behavior and won a wide following in the United States. McDougall believed in a teleological evolution, that is, a grand and intricate process with an ultimate purpose. His belief in the instinctual basis of social behavior was accepted by many American sociologists and economists between 1908 (when McDougall published his *Social Psychology*) and the 1920s. Yet a growing number of animal psychologists challenged the instinct theory. Their experiments showed that there was a much greater gap between higher primates and man than had been supposed. Only Watson, however, had conducted experiments in human instincts, and he had found no evidence to support the elaborate theories of instinctual behavior in human activity. Watson found his findings warmly received not only among psychologists but also among social scientists, who had, for some time, been uneasy with the instinct theory. Not only had the instinct theory's lack of experimental evidence been suspect, but those interested in the establishment of a scientific social psychology had stressed both the primacy of social and cultural environment rather than innate traits in accounting for the behavior of social groups and the importance of habit rather than instincts in the behavior of individuals.[25]

Moreover, reformers recoiled at the mounting racism and ethnocentrism during the 1920s. The resurgence of the Ku Klux Klan and the pressure on Congress to put ethnic quotas on immigration were only the most visible examples. The Princeton psychologist Carl C. Brigham set forth his views of Nordic superiority in *A Study of American Intelligence* (1923). Brigham's study, drawing heavily upon data gathered from the World War I army intelligence tests and reinforced with quotations from Madison Grant's notorious *The Passing of the Great Race*, gave the appearance of scientific sanction to those who lobbied for the Immigration Restriction Act of 1924. Brigham went on to become secretary of the College Entrance Examination Board, where he designed and developed the Scholastic Aptitude Test. Brigham's so-called scientific findings were heavily skewed against racial minorities and the disadvantaged, underscoring the conventional wisdom linking intelligence and race.[26] Behaviorism, then, with its emphasis on environment rather than on instincts, held special appeal for reformers. Not only did it invalidate assumptions of racial superiority, but it held out the hope that current racist attitudes could be changed through education and social conditioning.

Watson's experimental findings in 1917 had seemed to have sounded the death knell for the survival of the instinct theory, but in

1920 William McDougall came to Harvard's Department of Psychology. McDougall, with his ideas of "hormic energy," appeared to be reverting to the vitalism of the turn of the century. Furthermore, he offended progressive social scientists with his belief in Northern European racial superiority. Most alarming to American experimental psychologists, however, were his neo-Lamarckian notions of inherited behavioral traits and his reliance upon anecdotal rather than experimental evidence to support his theories. When McDougall published his *Outline of Psychology* in 1923, Watson was compelled to take the field as the self-appointed champion of American psychology. Watson wrote a stinging review for *The New Republic* in which he attacked McDougall's introduction of the notion of purpose into the study of human behavior. "McDougall's purpose," wrote Watson, "like Freud's 'unconscious,' Jung's 'libido' and Driesch's 'Entellichies,' is an insult to the corporate body of facts and deductions we call science."[27] Although most experimentalists would not have lumped Freud, Jung, and McDougall into one sweeping dismissal, there was general agreement with Watson's critique. Even E. B. Titchener, one of behaviorism's most vocal critics, admitted to a colleague that "I had a good deal rather be Watson than McDougall."[28] For to accept McDougall's opinions, it seemed, was to reject the whole direction of experimental psychology for the last quarter century.[29] McDougall, Watson argued, fostered a "lazy, genial, speculative, armchair attitude" toward psychology and reduced it to a prescientific status. "With purpose as a metaphysical presupposition," Watson maintained, "our behavior laboratories are useless."[30]

Indeed, the reaction to McDougall was so strong that it was to be another half-century before purposive behavior came under serious consideration in American psychology. This was, in large measure, due to Watson's finely honed skills as a propagandist. In 1924, the two men engaged in a public debate that was later published as *The Battle of Behaviorism*. McDougall's wit and erudition was pitted against Watson's polemics. Although Watson won the debate, it was a close call. Nevertheless, McDougall remained bitter toward behaviorism's visibility in the popular press and its prominence in academic circles. "Dr. Watson knows," he complained, "that if you want to sell your wares, you must assert very loudly, plainly, and frequently that they are the best on the market, ignore all criticism, and avoid all argument and all appeal to reason."[31]

Watson, however, was hardly one to avoid an argument. In fact, he courted controversy. Behaviorism needed a foil to give it shape and form. The growing interest in psychoanalysis during the 1920s gave Watson just such an opportunity. According to Watson, his "Freudian interests" had begun between 1912 and 1914, when his first wife announced that she had become "anaesthetic" toward him.[32] At the time, Watson's contact with Adolf Meyer at Johns Hopkins had put him in close touch with current trends in the psychoanalytic movement. In 1912, he had discussed his views with Meyer. On the one hand, Watson was hopeful. He believed that psychoanalysis added "another large territory to the province of psychology." The methods of Freud and Jung, he wrote, "contain within themselves the means of tremendously furthering medical practice, psychology and legal procedure." On the other hand, Watson was highly critical. Psycho-analysis, he insisted, had borrowed its methodology from psychology, and its devotees had formed a "cult" that had failed to maintain intellectual freedom and had hindered the scientific study of those methods.[33] By the 1920s psychoanalysis had won a popular following in the United States. Freudian terminology had become so ubiquitous, Watson complained, that "every shop girl will tell you of her dreams and complexes."[34] In 1923, Watson wrote to H. L. Mencken that he was "planning to 'assassinate' Jung's 'Psychological Types'." "In fact," he wrote, "I get more and more impelled towards assassinating the whole Freudian movement for its vagaries, in spite of the good that is in it."[35]

Watson's attitude toward Freudian psychology became increasingly focused upon a fundamental critique of the notion of the unconscious and upon what he considered to be the clinical imprecision of psychoanalysis. Following a successful series of lectures on "Behaviorism and Psychoanalysis" at The New School in 1926,[36] Watson was invited to speak at a symposium on the unconscious in Chicago with psychologist C. M. Child, Gestalt psychologist Kurt Koffka, and psychiatrist William Alanson White.[37] Watson argued that Freud's notion of the unconscious was based on a nineteenth-century concept of disease. Watson objected to the notion of mental disease or the "pathology of mind" because of the implication that "mind" was an entity distinct from the body. Reviving his earlier argument that thought was merely subvocal speech, Watson substituted "the unverbalized" in place of "the unconscious." Watson insisted that mental problems were actually *behavior* problems and that treatment should consist of a program of "unconditioning," or "retraining," rather than psychoanalysis. Successful psychoanalysis, he argued, actually involved a retraining process, but the procedure was imprecise, hit-or-miss, and unscientific. Treatment based on behavioral principles, Watson explained, could ignore vague concepts of diagnosis or cure and could institute a regimen of training that would "uncondition" deviant characteristics and inculcate patterns of desirable behavior.[38] In effect, Watson characterized psychoanalysis as, at best, imprecise and, at worse, ineffectual and wasteful. Behavioral methods, by contrast, were precise, efficient, and scientific. They could provide, Watson insisted, accurate clinical tools for engineering emotional health. Though he failed to convince many psychiatrists, Watson scored rhetorical points. He always played with an eye to the gallery. For those confused by the conflicting claims of rival psychologies, behaviorism seemed straightforward and practical. Watson's message did not go unnoticed.

In his review of *Psychologies of 1925*, a compilation of articles by leading psychologists, Ralph Barton Perry observed that "Perhaps the best proof that a branch of investigation has become a science is afforded by its ceasing to trouble itself about the matter. The doubtful sciences such as history, economics, sociology and ethics are those which are most insistent on being scientific. Psychology evidently still belongs to this list." According to Perry, what distinguished psychology from a true science was its "divided and conflicting program of research" and the lack of "an established technique and a body of generally accepted laws." It was evident to Perry that structuralism had passed from the scene. Although what had taken its place was not clear, he noted that "however much the authors of this book differ among themselves, there is an unmistakable likeness—at least of vocabulary. 'Behavior,' 'function,' 'process,' 'activity,' 'response,' these are now the terms to conjure by."[39] Perry's insight was shrewd. One manifestation of a consuming doubt is an obsession with certainty. Behavioral terminology provided one way for a "doubtful science" to redefine itself. Concrete, practical, seemingly scientific, behaviorism moved the investigation of human behavior to center stage.

Behaviorism's claims to demystify psychology and to simplify the complexities of modern life were also enormously appealing to middle-class Americans. Watson's ability to combine scientific jargon with the unadorned, direct style that he had refined as an advertising copywriter made him an effective propagandist. His first popular book, *Behaviorism*, was published in 1924 with a dedication to J. Walter Thompson president, Stanley Resor. This success prompted a series of articles in *Harper's*, which were compiled and published as *The Ways of Behaviorism* in 1928.

In these popularizations, Watson stressed that behaviorism was a psychology that was designed for modern living. The antecedents of behaviorism were not so much wrong as simply outmoded. The psychology of William James and its derivatives, as Watson put it, was "as much out of touch with modern psychology as the stage coach would be with modern New York's Fifth Avenue." The stage coach may have been "picturesque," but it had given way "to a more effective means of transportation." Watson could not have delivered a more stinging dismissal to his critics, for in the 1920s to be old-fashioned was to be irrelevant. In the machine age, no one wanted to be accused of living in the horse-and-buggy era.[40] Above all, Watson emphasized, behaviorism was not a spectator science. As he had proclaimed in 1913, Watson reiterated that "it is the business of behavioristic psychology to be able to predict and to control human activity."[41]

It was well that Watson had found his niche as a popularizer, for it was during this time that he lost the last tenuous foothold in academia that he had maintained at the New School for Social Research. Although the circumstances remain a mystery, Charles Beard's daughter recounted that in 1926 Watson's appointment was terminated at The New School because of "sexual misconduct." The charges were never specified or made public, but they were sufficiently serious to make Watson the first faculty member to be terminated since the founding of the New School.[42]

Though Watson's hopes for a return to academic respectability were dashed, he arguably had a much greater impact on psychology as a popularizer. More than any other pychologist of his generation, he shaped the image of the profession in the public mind. Moreover, his popularized vision of a science of behavior control stirred the imagination of a new generation of psychologists. It was a young B. F. Skinner who as a student glimpsed the "possibility of technological applications" in Watson's *Behaviorism* and devoured *The Psychological*

Care of Infant and Child in the aisle of a New York bookstore.[43] Indeed, much of Watson's popular writing envisioned a world where behavioral principles and methods governed the whole range of human conduct and social organization.

To justify such a massive reordering of society, Watson never tired of attacking fundamental social institutions. Marriage, the family, religion, and the law persisted because of "inertia and ignorance," he maintained. They were outmoded, old-fashioned, and fit for the "Victorian era". These institutions were already disintegrating, he believed. Indeed, he predicted that marriage would not survive the next fifty years.[44]

Watson had harbored many of these sentiments for some time. Even while at Johns Hopkins he had longed for a vehicle to influence public opinion on matters of personal conduct. In 1919, Watson had complained to Adolf Meyer:

> For the past two or three years I have been disgusted with the growing dominance of newspapers in controlling American life and I have often wondered why it was that scientific truths, facts of public interest and facts of public health gathered by scientists could not be presented to the public in such a way as to attract notice and become incorporated into our mores.[45]

Now that Watson himself had become an influential publicist, he used every opportunity to promote his vision of a world perfected by behavioral engineering. Yet the future that Watson revealed for his readers had an odd Victorian character. Like many of his contemporaries, Watson was, at best, ambivalent about modern life. Couched in futurist rhetoric, many of Watson's diatribes would have pleased the most reactionary antimodernist. On the other hand, some of his blueprints for social planning rivaled George Orwell's darkest nightmares.

In an article he titled "The Behaviorist's Utopia," Watson characterized the shortcomings of contemporary domestic life. The family, he declared, was "not geared for 20th century service." It had failed, he explained, "because it *prolongs the period of infancy.*"[46] On the frontier and on the farms, he argued, children had become "factors in production" at an early age. Even now it was not unusual, Watson wrote with approval, to see southern children of eight or nine at work in the cotton fields. But Watson blamed the very consumer culture he had helped bring about for eroding the integrity of the household. In the cities, he complained, children grow up without experiencing a struggle for survival and without becoming productive factors in the home. Housewives, Watson grumbled, were not only unproductive but could potentially subvert familial felicity. "Wives haven't enough to do today," he argued. "Scientific mass production has made their tasks so easy that they are over-burdened with time. They utilize this time," Watson concluded in his typically clamorous style, "in destroying the happiness of their children."[47]

Watson's invective against motherhood could not have been sustained in an era in which there was a consensus concerning family roles and functions. The more outrageous his statements, the more they indicate a period of flux and uncertainty in the American middle class. Watson believed that most mothers begin "to destroy the child the moment it is born."[48] Indeed, he dedicated his widely read *Psychological Care of the Infant and Child* to "the first mother who brings up a happy child."[49] Watson advised parents to treat their children as if they were "young adults." He outlined ideal parental behavior as "objective and kindly firm." This advice to parents was perhaps his most notorious:

> Never hug and kiss them, never let them sit on your lap. If you must, kiss them once on the forehead when they say good night. Shake hands with them in the morning. Give them a pat on the head if they have made an extraordinarily good job of a difficult task. Try it out. In a week's time you will find how easy it is to be perfectly objective with your child and at the same time kindly. You will be utterly ashamed of the mawkish, sentimental way you have been handling it.[50]

Watson's rationale for this controlled formality was his assertion that "all of the weaknesses, reserves, fears, cautions, and inferiorities of our parents are stamped into us with sledge hammer blows." Like Freud, Watson held that emotional disabilities were not inherited but inflicted upon children by their parents.[51] Few parents, Watson insisted, knew how to raise children; in his behaviorist's utopia, that margin of error would be eliminated. Watson had long dreamed of an experimental "baby farm" where hundreds of infants of diverse racial backgrounds would be the subjects of observation and research.[52] In his ideal world, child rearing would be brought as much as possible under laboratory control. Mothers would not know the identity of their children. Breast feeding would be prohibited, and children would be rotated among families at four-week intervals until the age of twenty.[53]

Watson justified such radical measures in light of his estimation of the "destructive" and "devastating" effects of what he called "mother love."[54] Yet his criticism of women in the workplace was equally severe. A life in the business world, warned Watson, made women unfit for matrimony. Watson believed that a career caused women to develop a "stratum of callousness and self-sufficiency" and "implants in her an artificial ambition to make a success in her work." In Watson's opinion, "a girl is foolish to spend her best years in the office . . . sharpening her brains when these things are of little importance to her in her emotional life as a woman."[55] Watson characterized women who dared to challenge the restrictions of their traditional social roles as maladjusted. "Does their demand for this mystical thing called freedom," he asked, "imply a resentment against child-bearing?" According to Watson:

> Most of the terrible women one must meet, women with the blatant views and voices, women who have to be noticed, who shoulder one about, who can't take life quietly, belong to this large percentage of women who have never made a sex adjustment.[56]

But if women were not fit for business, they were even less fit to be mothers. "No one today knows enough to raise a child," Watson complained. In fact, he suggested that the world would be better off if it ceased to have children for twenty years (except for "experimental purposes"), so that foolproof scientific methods of child rearing could be devised.[57]

In his ideal world, Watson would compensate for the shortcomings of wives and mothers by instituting a program of behavioral training. Women would be taught "domestic science," which covered everything from interior design to "the art of interesting and handling men" and the "technique of sex." Not only would young utopian women learn about the care and handling of children, but they would also study "the use of cosmetics, how to stay thin, how to be successful hostesses, and to put on the intellectual attainments that go into the making of a beautiful, graceful, wise woman." Indeed, all of the wives and mothers in Watson's utopia were beautiful and graceful because, as he chillingly put it, "large women and the occasional ill-favored woman are not allowed to breed."[58]

The function of the "biologically unfit" in Watson's world is unclear. Utopian social stability and harmony depended upon a rigid conformity. In such a society, the continual pursuit of banality absorbed energies that were potentially disruptive. In Watson's judgment, "the jobs of keeping themselves young and beautiful, useful, and in learning about home science gives [women] all the activity they need."[59]

These passages in which Watson's misogyny is so graphically revealed are some of the clearest examples of how thoroughly his personal prejudices pervaded his popular scientific writing. Yet Watson's attitude, extreme as it may have been, merely reflected an inherent contradiction in the emerging definition of the modern American woman. The attainment of women's suffrage in 1920 had diffused the thrust of organized feminism. With *normalcy* as its watchword, American society began a retreat from progressive politics in the 1920s, and the idea that the woman's place was in the home prevailed in conventional wisdom. But the home itself had radically changed. It was no longer set apart from commercial life—a "haven from the heartless world"—but was increasingly organized around the priorities of the marketplace. The home had ceased to be an enclave. The streamlined kitchen replaced the hearth, and efficiency rather than comfort dominated the functional designs of the modern house. The cadres of "experts" who barraged the homemaker with advice on the intricacies of managing "labor-saving" products were, by implication, testimony to the helplessness of the consumer.

In Watson's utopia, the implied authority of experts was institutionalized in the form of a technocracy managed by behavioral scientists. Like many futurist planners and social critics, Watson would have eliminated hereditary wealth from his new world. But Watson's rationale was far from egalitarian. Accumulated wealth and its associated power would be redistributed to the new elite of scientists with the object, as Watson put it, of obtaining "still further control over the psychological and physical environment." Watson's ideal citizens were neither communists nor capitalists, he argued; they were no longer guided by history or self-interest, but by science. Religion, which Watson saw as the antithesis of science, was to be outlawed. It encouraged "resignation, laziness and inefficiency" and excused "failure and weakness," he argued. Watson's new priesthood worshiped in the laboratory. All physicians, Watson insisted, would be trained behaviorists. Their incomes would rival that of corporate executives, and their decision-making powers would be unlimited. Preventive psychology would detect early "conduct deviation" in children. Any sign of "emotional upsets," truancy, or "unstandardized sex reactions" would warrant a reconditioning process that would eliminate "unsocial ways of behaving." Watson insisted that his ideal community would recognize no such words as "right, wrong or punishment." There would also be no mercy. When conditioning failed to cure what Watson termed the "hopelessly insane" or incurably diseased, the physician "would not hesitate to put them to death."[60]

Watson had tried to make a case for such draconian measures earlier in his best-selling book *Behaviorism* (1924). If the whole object of socialization was the production of useful individuals, Watson reasoned, then only "exaggerated sentimental and medieval religious mandates" prevented the liquidation of the "hopelessly insane." But social deviants and criminals, he argued, were merely "socially untrained." Those who failed to respond to reconditioning, he believed, could still be useful to society. "Failing to put on the training necessary to fit them to again enter society," he wrote, "they should be restrained always and made to earn their daily bread in vast manufacturing and agricultural institutions, escape from which is impossible."[61] Watson's panaceas for social instability could be dismissed as misanthropic rantings had not the twentieth century produced horrors that gave ghastly form to his nightmarish visions. Those who worshiped efficiency alone came to abandon concepts of good and evil, weighing their judgments on a scale that measured only degrees of order or disorder.

"Naturally," Watson pointed out, these measures implied the elimination of legal process. Watson looked forward to the day when "all law books are burned in some great upheaval of nature" and "all lawyers and jurists . . . decide to become behaviorists." Though legal procedures might be dispensed with, Watson noted that his system would by no means eliminate the need for "policing." But Watson reasoned that such enforcement would hardly be necessary, since his utopian citizens would be conditioned from birth.[62]

The success or failure of such a society depended upon the absolute control of an educational process that would function, not as a means of acquiring knowledge, but as the instrument of the individual's socialization. The whole purpose of education in Watson's perfected world would be to condition individuals to become completely absorbed in useful work. In fact, this was Watson's definition of "behavioristic happiness." Idleness or unproductive activity was to be anathema in the brave new behaviorist world. In Watson's view, "to be just an artist, just a poet, just a musician, or a composer, or actor calls for an examination at the hands of a group of utopian physicians." The citizens of Watson's utopia were trained from birth to function in a manner predetermined by a hierarchy of technocrats. Watson concluded that with such docile inhabitants, there would be "no need for that abstract entity we call the state."[63]

That Watson would consider such authoritarianism and regimentation to be the antithesis of a state reflects his assumption that a society based on scientific principles would be "non-political." But Watson, of course, had only banished the political *process* and its representative institutions from his ideal world. The fact that complete authority was to be wielded by scientists did not mitigate its totalitarian nature. Liberty had no place in Watson's vast anthill, where freedom was merely the freedom to be useful. Efficiency replaced justice in the arbitration of human activity. As in Kafka's dark world of *The Trial*, to be accused in a behavioristic society is to be guilty.[64]

The contradictions between individual freedom and social control inherent in behaviorism posed no dilemma for Watson. He left no doubt about where his sympathies lay. Watson admired the self-made man. "Are all the old buccaneers and pirates who used to make industry as interesting as war," he asked, "dead forever? While they made life unsafe for the weak," he declared, at least "they were builders."[65] Watson had nothing but contempt for those who championed social welfare legislation. Late in life, at the beginning of the McCarthy era, Watson heaped scorn on New Deal reforms. Franklin D. Roosevelt, he sneered, had made "democracy safe for himself, the non-voting population of the South, and for Mr. Stalin and his Communists."[66] Watson had no better opinion of Republican administrations. America, he wrote in 1930, was ruled by "professional politicians, labor propagandists, and religious persecutors." He characterized social experimentation from the beginning of human history as "infantile." Lacking an "educated ruling class," Watson lamented, society had been at the mercy of "some nation, political group, sect or individual, rather than under the guidance of social scientists. . . ." Watson's notion of "behavioristic freedom" seemed, on the surface, to imply the liberation of the individual from what he called the "taut steel bands" of "customs and traditions," but, in reality, Watson called for an imposition of rigid authority and the abolition of fundamental libertarian institutions. "I have always been very much amused by the advocates of free speech," he wrote. As Watson put it: "All true speech does stand substitute for bodily acts, hence organized society has just as little right to allow free speech as it has to allow free action." Speech and behavior alike could only be free to the extent that they did not run "afoul of group standards."[67] Watson clearly did not intend for behaviorism to imply that all human beings are equal, but rather that all human beings are *the same.*

Watson's preoccupation with social order and efficiency was one extreme of a broad spectrum of public opinion. The assumptions and values embodied in Watson's utopia reflected a "cult of efficiency" that dominated the thinking of American social planners from World War I through the New Deal. More than in any other period in American history, *good* and *efficient* came close to being synonymous. Not only did efficiency apply to the energy input–output ratio of a machine; it also came to stand for desirable personal attributes. An "efficient" person valued hard work over feeling discipline more than sympathy. Although these characteristics echoed the old small-town virtues of thrift, hard work, and self-discipline, in the twentieth century they could no longer be enforced by village consensus. In the modern world, efficiency and social harmony depended upon the "leadership of the competent." Moreover, competency itself came to be defined by a new class of professional managers.[68]

Watson's hierarchy of behavior technicians had their counterparts in the visionary literature of the progressive era. *Looking Backward*, Edward Bellamy's best-selling novel of 1888, had captured the imaginations of late–nineteenth-century reformers with its sentimentalized vision of a planned society. Mary Parker Follett, in her book *The New State* (1918), had argued for an "aristocracy of function" in a state that would be guided by the "new psychology" of Freud, behaviorism, and William McDougall's social psychology. Many future New Dealers were attracted to Thorstein Veblen's "technical alliance," which called for a "soviet of engineers" to effect a technical revolution that would replace an old order flawed by "conspicuous waste." Even John Dewey had suggested the establishment of an "intelligence trust" that would guide the destiny of those less endowed into successful "life adjustment."[69]

The concept of social efficiency provided progressives, who may have given lip service to democratic principles, with a basis for resisting the leveling tendencies of an emerging mass society. Calling for the establishment of an elite corps of engineers and planners to control the chaos of modern urban life, they sought to preserve the independence of the middle class by expanding the authority of professionals.

Watson continually sought to legitimize the authority of psychologists by emphasizing the application of their expertise to problems encountered in everyday life. Child care was considered to be especially fertile ground for behavioral engineering. In the right hands, he argued, the future of the next generation could be determined at will. "Give me a dozen healthy infants, well formed," read one of Watson's most quoted passages,

> and my own specified world to bring them up in and I'll guarantee to take any one at random and train him to become any type of specialist I might select—doctor, lawyer, artist, merchant-chief and yes, even beggar-man thief, regardless of his talents, penchants, tendencies, abilities, vocations, and race of his ancestors.[70]

Watson insisted that child rearing should be viewed as a "profession." Most parents failed miserably at bringing up well-adjusted children, Watson argued, because they were either unwilling or unequipped to assume the responsibility for such a rigorous task. The point that Watson drove home to mothers was that the fate of their children's well-being lay entirely in their hands. When a mother is first confronted by this thought, Watson wrote in *The Psychological Care of Infant and Child* (1928),

> she shies away from it as being too horrible. She would rather load this burden upon heredity, upon the Divine shoulder, or upon any shoulder other than her own. Once she faces it, accepts it and begins

to stagger under the load, she asks herself the question, "what shall I do? If I am responsible for what this tiny being is to become, where shall I find the light to guide my footsteps?" When such thoughts drive is it any wonder that there has been recently an almost frantic interest in what the laboratories of the behaviorist psychologists have to say about *infant culture?*[71]

Behaviorism stood ready to provide the guidelines for successful child development. Watson wanted to destroy at once the notion that each child carried within itself its own potential. A child was not like a seed that only needed nurturing to achieve the full flower of its own destiny. For Watson, a child was more like a lump of clay that needed to be molded, stamped, and hardened to become a useful vessel.

Watson offered advice on how to instill "neatness and order" and respect for the value of property. He discouraged what he called "sentimentality" in raising children. For Watson, a "happy child" is one "who never cries" unless physically injured; "who loses himself in work and play . . . who puts on such habits of politeness and neatness and cleanliness that adults are willing to be around him at least part of the day." Watson wanted to produce a child "who finally enters manhood so bulwarked with stable work and emotional habits that no adversity can quite overwhelm him." Watson made it clear that his goal was to produce children who would be able to cope with the realities of modern life. "I believe," he wrote, "that the internal structure of our American civilization is changing from top to bottom more rapidly and more fundamentally than most of us dream of." America's future was being manufactured in its factories and commercial institutions, and Watson wanted to insure that the next generation knew where to take their place on the assembly line of progress.[72]

Although Watson's book was intended for popular consumption, his ideas provoked responses from a wide range of critics on both sides of the Atlantic. Bertrand Russell admired Watson's attitude toward child rearing and tried to incorporate behavioristic principles in raising his own children. The results, according to Russell's daughter, were less than desirable.[73] In his review of Watson's book, Russell admitted that he did not wholly agree with Watson on all of his conclusions, but, for Russell, whether Watson was correct on all of his details was not important. Its "very great and rare merit" was that "its approach to the problem is scientific." For a positivist like Russell, the methodology was all-important. Watson was the exceptional writer to be commended to those interested in child care, because "he has studied the behavior of infants, not as parents study it, but as a man of science studies the behavior of amoeba."[74] What Russell did not realize was that so few of Watson's observations were based upon experimental evidence. Most of his conclusions seem to have been drawn from his belief that the future of society demanded the control he was seeking over human behavior.

Watson's concern with the social effects of parental affection was shared by many of his contemporaries. But unlike Sigmund Freud, John Dewey, and other theorists, Watson did not believe that affection would efficiently serve societal needs. Indeed, he argued that affection could potentially subvert the social order. He believed that psychological health required the complete adjustment of the individual to the needs of society. Watson, however, agreed with his contemporaries with regard to two important respects. First, he sought to professionalize child rearing by making parenthood an occupation requiring a familiarity with, or reliance upon, psychological expertise. His fondest wish would have been to bypass the family altogether and place the responsibility for child rearing entirely in the care of professional technicians. But perhaps the most significant effect of Watson's writing on the literature of the period was to reinforce the arguments of Freud and Dewey (and, later, Dr. Benjamin Spock) by emphasizing the importance of the early years of child development in the formation of adult personality and behavior.[75]

One of the most perceptive critiques of Watson's theories about child care came from Floyd Dell in his thoughtful study of *Love in the Machine Age* (1930). Watson's "revolt against mothering" and his attack on contemporary family life was seen as a jump "from the patriarchal frying-pan into the patriarchal fire," since the behaviorist's substitution for traditional family authority was the technocratic authority of psychological engineers. For Dell, Watson protested too much in his hatred of what was actually neurotic mothering. He perceptively noted in Watson's indignation an "almost broken-hearted defense of children against the evils of 'coddling.'" But Watson went too far, said Dell, when he advised against the demonstration of genuine parental affection. Citing studies that indicated that the withholding of affection produced lasting and socially destructive damage, Dell maintained that the strictures advocated by Watson were actually beside the point when it came to intelligent child-training. "Psychologically healthy parents," argued Dell, "will be able to break all of Dr. Watson's rules with impunity," just as "every one of these rules may be strictly obeyed by neurotic parents with neurotic results." The issues for Dell were those of freedom and control. "Children can be emotionally exploited," he wrote, not only in the name of "Religion, of Duty, or Morality, and of Love," but also in the name of "Science, of Freedom, of Freud, and of Dr. Watson." Dell did find many things in Watson's work to admire. For him, behaviorist techniques could, in the hands of healthy parents, provide useful means for raising children who were sure of their parents' love for them. But since Watsonian man was a machine, the mechanistic means could not be separated from the goals of regimentation and control. With such limited horizons, argued Dell, both parents and children would be enslaved by a technique and a methodology ironically dedicated to their happiness.[76]

Thoughtful critics like Dell, however, were overshadowed by Watson's showmanship. Watson spread the behaviorist faith with a zeal matched by few of his contemporaries. A sensation-hungry press competing for a share of the mass-circulation market happily offered a forum for Watson's opinions. His name became a familiar by-line in such magazines as *Harper's*, *The Nation*, *The New Republic*, and *The Saturday Review of Literature*, as well as in *McCall's* and *Liberty*. Profiled in *The New Yorker*, he was the subject of articles and interviews in countless newspapers and Sunday supplements as reporters learned that he could be counted upon for "good copy" on almost any topic. Watson commented on the future of marriage, the origin of fears and aversions, and the usefulness of psychology in prisons.[77]

Though Watson's writing was prolific, he showed a particular lack of enthusiasm for literature. Yet he never hesitated to offer his strong opinions on the subject. In an article for *The Saturday Review* called "Feed Me on Facts," Watson took novelists, playwrights, artists, and critics to task for using their imagination rather than the observation of human behavior as the basis for their creations. "Before I die, I want to see one good author and one good playwright get a behavioristic background," he wrote. "I want him to take man as a piece of squirming protoplasm and make him interesting—without overstepping the bounds of natural science in doing it." Watson had nothing but contempt for what he referred to as the "psychological novel." D. H. Lawrence was singled out for special condemnation. His main objection was that authors like Lawrence were "stupidly ignorant" of psychology—of course, by psychology Watson meant behaviorism. There was no need to fall back on the "claptrap" of sentimentality and romance, he argued. A character's motivation should be based on environmental conditioning. Whatever literary gifts Watson may have had, he totally lacked a sense of irony. "There are no mysteries in real life," he wrote, "why can't we keep them out of books?"[78]

Watson may have liked his fiction hard-boiled, but he was only expressing crudely what many writers of his generation had come to believe. The tradition of American realism, which had begun with Twain and continued through Stephen Crane, had received perhaps its ultimate expression in what one critic called the "stimulus-response behavioristic fashion so characteristic of Hemingway."[79] In 1919, T. S. Eliot had advanced his notion of the "objective correlative" by which emotion could be expressed in poetry only by "finding . . . a set of objects, a situation, a chain of events . . . such that when the external facts, which must terminate in sensory experience, are given, the emotion is immediately evoked." After all, William Carlos Williams remarked, "a poem is a . . . machine made of words."[80] As Cecelia Tichi has recently shown in *Shifting Gears: Technology, Literature and Culture in Modernist America*, for the modernist artist and writer, the machine became a powerful metaphor that connected human beings to their world.[81] The Dynamo had indeed replaced the Virgin, but what would keep it from propelling the world into fragments? It was preoccupation with control that permeated every level of American life and culture during that period. Like Watson, twentieth-century writers struggled, in their lives and in their work, to make sense out of the confusion and dislocation of modern life. But if the trend in American fiction was toward the kind of "realism" that Watson admired, some critics lamented the modern world that it reflected. Writing in 1922, Van Wyck Brooks shrewdly observed that "our novelists show us that in order to succeed in life one cannot be up and doing too soon. The whole temper of our society, if one is to judge by these documents, is to hustle the American out of his childhood, teaching him at no age at all how to repel life and get the best of it and build up the defenses behind which he is going to fight for his place in the sun." This attitude might "produce sharp-witted men of busi-ness," Brooks admitted, but, he asked, "could anything be deadlier to the poet, the artist, the writer?"[82]

For Watson, the true measure of artistic merit lay in the market-place. "If the individual is a writer," Watson proposed,

> we should want to draw a curve of the prices he gets for his stories year by year. If from our leading magazines he receives the same average price per word for his stories at 30 that he received when he was 24, the chances are he is a hack writer and will never be anything but that.[83]

By his own standards, Watson was a very successful writer indeed. He complained about being besieged by newspapers for inter-views, and he received a steady stream of proposals for books and articles from the likes of V. F. Claverton and H. L. Mencken.[84] As his reading audience grew, he also became a familiar voice to a vast listening public. When radio broadcasting became a national industry in the 1920s, Watson took his behaviorist message to the airways. On one nationwide broadcast for the National Broadcasting Company, Watson told his listeners "how to grow a personality." He described how the control of a child's environment was essential to the control of the development of its personality and held out the hope that future societies would have the ability to produce the personalities they required on demand.[85] As crude as his message was, it did not fail to make an impact as behaviorism was beamed into the nation's living rooms.

A generation mystified and apprehensive about the secrets of science turned with relief to a popularizer like Watson. On the whole, Watson's popular writing was well received by critics. Indeed *Behavior-ism*, published in 1924, had been hailed by *The New York Times* as initiating "a new epoch in the intellectual history of man."[86] Watson's writing was described by the *Atlantic* as "more revolutionary than Darwin, bolder than Nietzsche, and . . . more useful to the human race than the fatalistic eugenicist."[87] Culture critic Stuart Chase was moved to describe *Behaviorism* as "perhaps . . . the most important book ever written. One stands for an instant blinded with a great hope."[88]

Not all of Watson's readers looked upon the world he described with equanimity. Some contemporaries in the antimodernist tradition despaired over the enthronement of scientific materialism and the displacement of traditional values.[89] Others challenged behaviorism on its own ground. Journalist Harvey Wickham called Watson and his devotees to account in *The Misbehaviorists*. If science in general, and behaviorism in particular, truly represented the world as it was, Wickham argued, he was ready to accept it. But Wickham maintained that scientists themselves were neither in agreement nor logically consistent in the presentation of their world views. For instance, Wickham found the denial of consciousness acceptable for the observation of behavior in the laboratory but unfounded and unsupportable when applied to generalized assumptions about the nature of man. Similarly, if Watson believed that the three basic emotions were fear, rage, and love, his use of fear and rage but exclusion of love in conditioning children was based on his perceptions of social needs rather than scientific evidence.[90]

Watson had tirelessly preached that science or a positivistic world view would ultimately replace religion in shaping the direction of human aspiration. Yet his own extreme materialism came under attack for the same reasons that he had attacked religious belief.[91] Mortimer Adler despaired that "Watsonism has become gospel and catechism in the nurseries and drawing rooms of America." For Adler, Watson's scientific work had degenerated into "generalized bigotries." Watson, Adler complained, had become "the exponent of his own evangelical religion and he preaches it with all of the dogmatic zest and vulgarity of Billy Sunday."[92] Watson had clearly gone beyond his own facts in promoting behaviorism. It is ironic that such a dedicated apostle of scientific materialism, who had protested so vehemently against the kind of evangelical religiosity that pervaded his own childhood, would be accused of the same excess of zeal in advancing his own ideas. Yet Watson was so steadfastly confident in the inevitable proof of his theories that he demanded a leap of faith on the part of those who would call themselves behaviorists. For the true believers, Watson promised a heavenly city on earth, for it was to the future that Watson looked for salvation. The success or failure of behaviorism rested on the results of its application to future generations of children. Thus it was posterity that would feel its blessing or its curse.

In 1930, an evaluation of behaviorism by philosopher Horace Kallen appeared in the first edition of the *Encyclopedia of the Social*

Sciences. Kallen credited Watson with having brought about an "intellectual revolution." Like psychoanalysis, it "reoriented the mind with respect to the nature of man." Behaviorism was seen to be the result of a "purely theoretic urge toward mechanistic precision . . . reinforced by a practical one deriving from the industrialization of American life." It said with authority what many in America had already come to believe. It made adjustment to the world simpler and human nature more hopeful. Watson, wrote Kallen, made human infants at birth, regardless of their heredity, "as equal as Fords." But along with the democratic assumptions of that mechanistic metaphor came the darker implications of assembly line production and interchangeable parts. Behaviorism, wrote Kallen, gave the idea of social control "a sure ground."[93]

By the end of the decade, a major appraisal of the behavioristic movement in the *Journal of General Psychology* credited Watson with bringing psychology into "harmony with the universal method of science."[94] Watson was praised for recognizing the "practical need" for the development of an objective methodology. Behaviorism was "tough-minded," but it also promised to perform miracles. In Watson's hands it became "a moral faith in science, 'a religion to take the place of religion,'" and a faith that stirred the imagination and commanded the loyalties of an entire generation of younger psychologists. Not only did Watson alter the methodology and terminology of the science of psychology, but even the "blunderbuss polemical methods" of his popular writings had the effect of making "psychology in America a vastly more colorful and interesting calling."[95] Watson understood the importance of roles and "images" in a society in which mobility had uprooted traditional familial and community ties. He clearly saw the possibilities in the emerging science of psychology both as a vehicle of social mobility for a rising professional class and as a means of providing direct services of social control for an emerging corporate society that sought stability and predictability. For, as Horace Kallen commented on behaviorism, "the life of a doctrine, like the life of a man, varies with its bearing on the ruling passions, the dominant drives of the community in which it appears."[96]

Watson's utopian vision was not mere hucksterism—it was part of the progressive dream. It spoke to an era in which science was to become a new religion and the scientific method was to create a binding faith for its practitioners. It was, above all, a faith in a radical environmentalism that invested man with the ability to make and shape his own world, free from the authority of tradition and the dead hand of the past. A "nation of villagers" had found themselves in a society that had rapidly become urbanized and industrialized. Rapid change was the order of the day, and the happiness or misery of humanity depended on a willingness to bring the process of change under control. But if behaviorism represented the freedom to remake the individual, it also posed the possibility of directing human activity into predetermined channels. It was the latter aspect of behaviorism that Watson chose to emphasize. The institutionalization of progressive policies gave rise to new forms of social control that stabilized the industrial order without solving any of its underlying problems. The emerging technocracy was able to keep social tensions from taking political form but was not able to remove their sources. In solving human problems Watson's impersonal characterization of man as machine held tremendous appeal for the managers of a mass society. For as envisioned by Watson, behaviorism was to serve the authority of those who desired a stable and predictable social order.[97]

10

Epilogue

When you're dead, you're all dead.
JOHN B. WATSON

In 1932, Robert Yerkes encouraged Watson to "bury advertising and return to observational work." "Surely," he quipped, "you can sufficiently forget your behavioristic philosophy to be happier in experimentation than in generalization."[1] Watson's reply was equally barbed. My "habits and tastes," he retorted, are "geared far beyond the return I could hope for from any academic job, even assuming that any university in the country were so misguided as to offer me a job."[2] In the dismal midst of the Great Depression, when many lives were thrown out of gear, Watson's mechanistic metaphor was doubly pointed. Watson, however, reveled in all the outward signs of achievement and prosperity. For all appearances his public life symbolized the American dream of success. An internationally acclaimed scientist, popular publicist, and successful businessman, he seemed to commute between the two worlds of his Connecticut estate and Madison Avenue with ease and confidence. Yet nothing was obvious about John B. Watson except the masks he chose to wear, obscuring from everyone, including perhaps himself, his true personality.

He was a man who loved to race speedboats at full throttle on Long Island Sound but refused to drive an automobile. He had a taste for fine clothes and the company of New York society, but he also prided himself on being able to take on all challengers in extensive drinking bouts. Even then, it was merely the appearance of abandon, for he maintained control by downing shots of olive oil beforehand to prevent the alcohol's absorption. He had few friends but was obsessed with the pursuit of women. One of his business colleagues expressed wonderment that anyone as "hedonistic in his personal philosophy could in his early years have been such a rigidly disciplined experimental scientist." As an advertising executive, he was the model of an urbane New York sophisticate, but he was uncomfortable with the pace of city life. He built a rural paradise in southern Connecticut. There, in his Tudor-style manor house, he lived the life of a country squire. But even there, Watson was ambivalent about the life he had created for himself. While he had his shirts and shoes hand-made in England, on occasion, attired in overalls and a plaid shirt, he delighted in mocking his guests with rough-hewn manners and backwoods hospitality. Though he liked to boast casually about his retinue of servants, he also loved to do much of the labor on his estate himself. Building his barns and his piggery, taking care of his cattle and horses were favorite pastimes. Echoing Victorian schoolmasters who once admonished their charges to "obey the rules and lose one's self in work," Watson sought happiness in self-absorption. As he put it: "The only freedom worth striving for is complete engrossment in activity."[3]

Yet only certain activities provided Watson with happiness. He could spend long, tedious hours in the laboratory or working on his estate but could rarely force himself to sit through a play. Watson's youngest son later recalled that his father would bolt from the room if conversation changed from the casual to "deep" subjects. There was always work to be done in the garden, or the dogs (Watson's favorite companions) demanded immediate attention. Watson's preoccupation with being busy suggests something other than a search for pleasure; his constant mechanical motion more resembles a flight into the *oblivion* of activity. One gets the sense of sheer panic, barely suppressed. What Watson sought most desperately to escape can only be guessed at. He steadfastly refused to reflect upon his own life. His scant autobiographical writings are curiously flat and omit much more than they reveal. Watson feared that a survey of one's past could lead only to despair. "I don't see how anyone except a very naive person could write up his own life," he wrote.

> Everyone has entirely too much to conceal to write an honest [autobiography]—too much he has never learned to put into words even if he would conceal nothing. Think of chronicling your adoles-

cents acts day by day—your four years of college—your selfish-
ness—the way you treat other people—your pettiness—your day
dreams of sex! Autobiographies are written either to sell the good
points about oneself or to vanquish one's critics. If an autobi-
ographer honestly turned himself inside out day by day for six
months, he would either commit suicide at the end of the time or
else go into a blissful oblivescent depression.[4]

Evidently Watson believed that only the unexamined life was
worth living. As if he were trying to convince himself, Watson (in one
of his few unpublished articles) elaborated on the theme of "why I
don't commit suicide." Watson wrote to his former colleagues in
psychiatry and psychology, ostensibly to gather data for his article,
but perhaps seeking reassurance. The conclusion that he drew is
revealing. "The best, sanest and quickest acting thought I can give to
one brooding over taking his own life," Watson wrote, "is this: Run
away for a week, a month or a year. There is no psychological
medicine so potent in this wide world as a *new environment*."[5] In
America, where "heading west" was no longer a way of solving one's
problems, Watson's advice had particular significance. As one of the
architects of modern advertising, Watson understood the magic of the
word *escape*. The successful marketing of mass entertainment and a
vast array of consumer goods depended on the perpetual state of
anxiety produced by the pressures of modern living. That Watson
himself was a victim of those pressures is not surprising, and his
response was typical of many Americans. His pride in exhibiting a
"grace under pressure" was reflected in the code of the strong, silent
hero in the popular film and fiction of the period. Yet Watson's
inability or refusal to face the welter of confusion and contradictory
impulses that dominated his actions forced him into a rigid, desperately one-dimensional view of life that could tolerate no ambiguity.
What many took to be callousness or indifference was, in reality, an
extreme sensitivity to the uncertainties of daily existence.

Nevertheless, the consequences of such behavior were often
tragic. Watson's defensive preoccupation with self-control frequently
blinded him to the sensibilities of those closest to him. The ads Watson
used to promote *The Psychological Care of Infant and Child*, for instance,
included photographs of his two sons, who were described as "Behavioristic Children . . . who [*sic*] adults like to be around. They lose

themselves in work or play. They sleep and rest when put to bed . . . they eat what is put before them. They are free from fears and temper tantrums—*they are happy children.*"[6] But, in fact, their childhood was far from happy. Watson's youngest son later remarked that growing up with his father was "like a business proposition." Their relationship was "devoid of emotional interchange," but the children were ex- pected to be extremely meticulous in their bodily habits and punctual at meals and at bedtime.[7] Though Watson did not inflict a full-fledged behavioristic regimen on his children, he was not above testing their reactions from time to time. In one heart-breaking passage from *Behaviorism*, Watson described how he subjected his eldest son, Billy (then about three years old), to an experiment to determine his instinct for jealousy by appearing to physically abuse his wife in front of the child. Terrified and confused, Billy "cried, kicked and tugged at his father's leg and struck with his hand." Yet Watson continued the display of violence until "the youngster was genuinely disturbed and the experiment had to be discontinued."[8]

The distance that Watson put between himself and his children was never more apparent than when his wife, Rosalie, died suddenly from pneumonia in 1935. The night their mother died, the children were sent away to camp and brought home the following morning. Watson's youngest son recalled that he and his brother learned of their mother's death from the cook. It was the only time that he remem- bered seeing his father cry. Standing in the doorway in front of the house, Watson put his arms lightly around the shoulders of his sons. That moment was the one occasion in which Watson's children experienced a genuine expression of intimacy and affection from their father. Watson never discussed his wife after that.[9]

Rosalie had been a "vibrant, sparky, fun-loving person" who loved the theater, parties, and an active social life. She had the ability to draw Watson out of his aloof, almost shy disposition. Watson distanced himself even more from the world after her death. He began drinking heavily and plunged into a routine of work, at the office and on his estate. The social aspects of his life all but disappeared. He became much less inclined to experiment with new friends or, for that matter, new ideas. He avoided company, but when company appeared he used his charm and his sense of humor to parry any serious turns in the conversation. This period was especially difficult for the children and particularly hard for Watson's eldest son. As he grew older, Billy passed through a restless adolescence and often quarreled bitterly with his father. The estrangement deepened when, following graduation from college, he decided to become a psychiatrist. Rather than encourage his son's decision, Watson took it as a slap in the face. An uneasy peace ensued between the two, but their conflicts were never resolved. The circumstances of that relationship were unfortunate enough, but the ultimate tragedy came within four years of Watson's death, when his son took his own life.[10]

Watson left the J. Walter Thompson Company in 1935 and became an advertising executive with the William Esty Company, where he remained until his retirement in 1945. During this period he became even more of a recluse. His granddaughter recalled visiting "Big John," who lived alone in the empty house that always seemed to smell of bourbon.[11] There he kept busy with a constant series of projects. Although a number of people worked on the farm, he never asked others to join his activities, preferring to remain alone. At night he would be up until the small hours caring for his dogs or reading boxloads of Western novels and detective stories.[12] Watson's choice of reading material reflects the deep divisions within himself that he shared with many other Americans. For the detective novel represents an attempt to control the tide of modern chaos and unrestrained passion by the application of order and reason, while the Western hero exists in a preindustrial world where survival depends on action rather than reflection.

It was perhaps an attempt to resolve that conflict that led him to sell his estate after World War II and move to the hills of western Connecticut, where he spent the last years of his life in a small, frame farmhouse that bore a striking resemblance to his boyhood home in South Carolina. There, with his secretary-companion, he found a certain peace in the rituals of growing fruit trees and caring for the animals on the farm. In the winters he busied himself in a woodworking shop that he kept in the barn, where he slept during the summers.[13] If this "complete absorption in activity" provided him with a measure of "behavioristic happiness," it was not enough to fill the void in his life. In 1954, his former protégé, Karl Lashley, wrote to E. G. Boring that he had heard from Watson, "mentioning that he hopes to get back to whiskey soon, from which I judge that he has had to cut out his usual quart per day."[14]

In his self-imposed isolation, Watson had all but cut himself off from those former colleagues who still remembered him. But in the

fall of 1957, the American Psychological Association planned to honor
Watson at their annual meeting in New York. Watson was almost
eighty, and to most members of the Association he was a legend. His
secretary drove him down to the city on the day of the meeting, but at
the last minute Watson refused to go inside and insisted that his eldest
son attend in his stead. After almost forty years of exile, Watson was
to be publicly recognized by the profession he had helped shape for the
"revolution in psychological thought" that was one of the "vital
determinants in the form and substance of modern psychology." But
Watson was afraid that in that moment his emotions would overwhelm him, that the apostle of behavior control would break down
and weep.[15]

Watson died "peacefully" a little more than a year later. He was
eighty years old. But he managed to accomplish one more task before
he died. Gathering up a lifetime of correspondence, research notes,
and manuscripts, he carried them to the fireplace of his clapboard
farmhouse and slowly burned them one by one. When his secretary
protested the loss to posterity and to history, Watson only replied:
"When you're dead, you're all dead."[16]

Watson had come a long way from Reedy River, but at the end
of the road he had found, after all, only himself. Turning his back on
the world he had helped design, he lived out the end of his days as he
began them: to the rhythm of daily farmyard ritual. Perhaps Watson
endeavored to preserve an enclave of self-reliance in a society where
it no longer made sense. If this was the case, he was hardly alone. By
1957, millions of surburban outposts, complete with "ranch houses,"
had sprouted on the frontiers of cities. During the same period Walt
Disney's dream factory had transformed Davy Crockett into a cultural hero and enshrined him in a fantasized "Frontierland" that
sprawled alongside a southern California freeway. The success of this
type of commodified sentimentality was one of the outward signs of
the triumph of a corporate culture. For it pandered to nostalgia for a
lost social order that it had itself destroyed.[17]

Watson was certainly among those who called the loudest for the
demolition of late–nineteenth-century social institutions. He was a
key figure in the ranks of those who facilitated the transformation of
American life from an individualistic to a corporate society. By
shaking off the restraints of Victorian custom and tradition, they
hoped to release and harness the energies at the heart of the emergent

modern culture. Behaviorism promised to unlock the mechanism that governed human action. But its scientific foundations were based on an already outmoded nineteenth-century positivism. Nevertheless, behaviorism's notion of a malleable human nature fired the hopes of a new meritocratic elite. Watson emerges as a figure linking this class and the cult of efficiency with the rise of a national advertising industry and the gospel of self-fulfillment. For Watson shared the delusions characteristic of many twentieth-century planners: that efficiency and exhilaration were ends in themselves—a conception common among those who could no longer envision social or ethical goals.[18] What Watson had in common with many of his contemporaries was a preoccupation with control. In Watson's case, however, it was an obsession. Watson's singleminded pursuit of behavioral control blinded him to the ways in which the human species actually behave as social beings. If Watson had been more interested in the pursuit of knowledge, he might well have heeded the words of William James. In 1890, two years before the American Psychological Association was organized, James observed that psychology had not yet produced its Galileo and its Lavoisier. "Meanwhile," he wrote,

> the best way in which we can facilitate their advent is to understand how great is the darkness in which we grope, and never to forget that the natural-science assumptions with which we started are provisional and revisable things.[19]

Notes

CHAPTER ONE

1. John B. Watson to William Rainey Harper, July 20, 1900, Presidents' Papers 1889–1925, University of Chicago Archives, Chicago, Ill., cited hereafter as Presidents' Papers.

2. See: Burton Bledstein, *The Culture of Professionalism: The Middle Class and the Development of Higher Education in America* (New York: W. W. Norton, 1976).

3. Genealogical information on the Watson family, compiled by Mr. and Mrs. J. N. Watson, is on file in the Furman University Library, Greenville, S.C. Additional information was furnished by Thomas Roe of Greenville, S.C. In addition, David Cohen's *John B. Watson, The Founder of Behaviorism: A Biography* (London: Routledge & Kegan Paul, 1979) should be noted. Its numerous inaccuracies and anecdotal style, however, make it an unreliable, if sometimes entertaining, source.

4. Mildred W. Goodlett, *The History of Travelers Rest* (Greenville, S.C.: Author, 1966), p. 2.

5. Guy A. Gullick, "Greenville County: Economic and Social," *University of South Carolina Bulletin* CII (September 1921): 7–9; Harvey T. Cook, *Education in South Carolina under Baptist Control* (Greenville, S.C., 1912), pp. 164–65; Goodlett, p. 3.

6. Pickens Butler Watson, Jr., to Myrtle B. McGraw, August 22, 1943, courtesy of Mrs. Ruth Young (niece of John B. Watson).

7. Ibid.

8. Ruth Young, telephone interview, May 16, 1976; Dr. Ben Field, quoted in: Charles L. Brewer, "John B. Watson, Product of Furman University" (Paper presented as part of a symposium on "History of Psychology in the South" at the Annual Meeting of the Southern Society for Philosophy and Psychology, Atlanta, Ga., April 15–17, 1976), p. 4. Additional information on the Watson family was obtained through interviews with: James B. Watson, Watson's son by his second marriage (interview held on July 11, 1982, Palos Verdes Estates, Calif.) and with Mary Watson Hartley, Watson's daughter by his first marriage (interview held on July 12, 1982, Los Angeles, Calif.).

9. Hartley, interview.

10. William Mann Batson, *History of Reedy River Baptist Church* (Greenville, S. C.: Poinsett Printing Company, 1959), p. 28.

11. John B. Watson, in *A History of Psychology in Autobiography*, Vol. III, ed. Carl Murchison, (Worcester, Mass.: Clark University Press, 1936), p. 271; Watson and Hartley interviews. Watson was named for John Albert Broadus, a prominent educator and theologian at the Southern Baptist Theological Seminary. Broadus left Greenville when the seminary moved to Louisville, Ky., in 1877, a year before Watson's birth. For an account of Broadus's impact on Southern Baptists, see: Lucille T. Birnbaum, "Behaviorism: John Broadus Watson and American Social Thought, 1913–1933" (Ph.D. diss., University of California at Berkeley, 1964), pp. 15–23. See also: John B. Watson and Will Durant, "Is Man a Machine? A Socratic Dialogue," *The Forum* LXXXII (November 1929): 264. Late in life, Watson wrote to E. G. Boring that he was "named after old doc John A. Broadus, the Great Baptist divine—who was a friend of my grandmother Roe (platonic, so far as I know)." See: Watson to Boring, April 1, 1955, E. G. Boring Papers, Harvard University Archives, Cambridge, Mass.

12. See: Philip Young, *Ernest Hemingway: A Reconsideration* (New York: Harcourt, 1966), pp. 188–92.

13. Watson, *Autobiography*, pp. 271–72.

14. Ibid; Pickens Butler Watson, Jr., to McGraw; Goodlett, p. 18.

15. Pickens Butler Watson, Jr., to McGraw; Young, telephone interview.

16. Between 1870 and 1890 the number of adult males employed in manufacturing industries in South Carolina doubled. The base of this expansion was the cotton textile industry. In the same era the number of textile employees increased by six times, while the number of factories tripled and the productive capacity multiplied tenfold. As the industry demonstrated its ability to absorb the setbacks of the panics of the 1870s and 1880s, it began to attract national attention and the interests of northern capital. See: U.S. Department of the Interior, Census Office, *Population of the United States at the Eleventh Census: 1890*, Vol. I, pt. 2 (Washington, D.C.: U.S. Government Printing Office, 1895), p. 335; U.S. Department of the Interior, Census Office, *Report on Manufacturing in the United States at the Eleventh Census: 1890*, Vol. VI, pt. 3 (Washington, D.C.: U.S. Government Printing Office, 1895), pp. 188–89; "South Carolina's Prosperity," *New York Times*, February 4, 1884, p. 1, c. 3.

17. Broadus Mitchell, *The Rise of Cotton Mills in the South* (Baltimore: The Johns Hopkins Press, 1921), p. 173; U.S. Department of the Interior, Census Office, *Population of the United States at the Ninth Census: 1870*, Vol. I. pt. 1 (Washington, D.C.: U.S. Government Printing Office, 1872), p. 259; U.S. Department of the Interior, Census Office, *Population of the United States*

at the Eleventh Census: 1890, Vol. I, pt. 1 (Washington, D.C.: U.S. Government Printing Office, 1895), p. 308; U.S. Department of the Interior, Census Office, *Population of the United States at the Twelfth Census: 1900*, Vol. I, pt. 1 (Washington, D.C.: U.S. Government Printing Office, 1901), p. 474.

18. Broadus Mitchell and George Sinclair Mitchell, *The Industrial Revolution in the South* (Baltimore: The Johns Hopkins Press, 1930), pp. 9–10; Ernest McPherson Lander, Jr., *A History of South Carolina, 1865–1960* (Columbia: University of South Carolina Press, 1970), pp. 84–85.

19. Mitchell and Mitchell, pp. 11, 32–33.

20. Lander, p. 30.

21. Albert N. Sanders, "Greenville and the Southern Tradition," *The Arts in Greenville, 1800-1860*, ed. Alfred S. Reid (Greenville, S.C.: Furman University Press, 1960), pp. 136–37; Alfred S. Reid, "Literary Culture in Mid-Nineteenth Century Greenville," *Proceedings of the Greenville County Historical Society, 1962–1964* (Greenville, S.C., 1965), p. 74.

22. S. S. Crittenden, *The Greenville Century Book* (Greenville, S.C.: Greenville *News* Press, 1903), p. 14.

23. Laura Smith Ebaugh, "A Social History," *Arts in Greenville*, ed. Reid, pp. 18–22.

24. See: Robert H. Wiebe, *The Search for Order: 1877–1920* (New York: Hill & Wang, 1967), pp. 118–19.

25. Watson, *Autobiography*, p. 271; Watson to Robert M. Yerkes, July 17, 1910, Robert M. Yerkes Papers, Yale University Medical Library, New Haven, Conn., cited hereafter as Yerkes Papers.

26. Lander, p. 29; George Brown Tindall, *South Carolina Negroes, 1877–1900* (Columbia: University of South Carolina Press, 1952), pp. 235, 238–39; "The Spartanburg Lynching," *New York Times*, July 21, 1879, p. 5, c. 3.

27. Robert Norman Daniel, *Furman University, A History* (Greenville, S.C.: Furman University Press, 1951), p. 76; "To Suppress Race Wars," *New York Times*, November 18, 1898, p. 1, c. 7; Mary Wyche Burgess, "John Broadus Watson, 1878–1958" typescript, Greenville, S.C., March 1973, p. 5; Crittenden, p. 61.

28. Laura Smith Ebaugh, *Bridging the Gap: A Guide to Early Greenville, South Carolina* (Greenville, S.C.: Greenville Chamber of Commerce, 1966), p. 7; Crittenden, p. 57.

29. Crittenden, p. 60; Robert C. Tucker to Lucille T. Birnbaum, March 9, 1961, Baptist Historical Collection, Furman University, Greenville, S.C., cited hereafter as Furman University.

30. Tucker to Birnbaum, February 21, 1961, Furman University.

31. Louis Berman, *The Religion Called Behaviorism* (New York: Boni & Liveright, 1927); Tucker to Birnbaum, February 21, 1961, March 9, 1961, Furman University; Young, telephone interview; for a speculative, psychohistorical analysis of Watson's religious background as an influence in the

development of behaviorism, see: Paul G. Creelan, "Watsonian Behaviorism and the Calvinist Conscience," *Journal of the History of the Behavioral Sciences* X (January 1974): 95–118.

32. Page Smith, *As a City upon a Hill: The Town in American History* (New York: Alfred A. Knopf, 1966), pp. 238, 251–52.

33. Lander, p. 130; Watson, *Autobiography*, p. 271.

34. Smith, p. 242; R. H. Knapp and H. B. Goodrich, *Origins of American Scientists* (Chicago: University of Chicago Press, 1952), p. 293; Bledstein, Chapters 6–8.

35. Cook, p. 57; Daniel, pp. 20, 44–49, 58.

36. Cook, pp. 152, 122. The state university at Columbia during this era was hardly a hotbed of free thought. One professor was removed from the faculty because of his religious views. The professor of logic (a former Baptist clergyman) had declared himself to be a Unitarian. See: "No Unitarians Wanted," *New York Times*, May 15, 1891, p. 1, c. 3.

37. Sanders, pp. 136–37.

38. W. J. McGlothlin, *Baptist Beginnings in Education, A History of Furman University* (Nashville, Tenn.: Southern Baptist Convention, 1926), pp. 151–69; see also: Bledstein, pp. 203–286; and Richard Hofstadter, "The Age of the University," *The Development and Scope of Higher Education in the United States*, eds. Richard Hofstadter and C. DeWitt Hardy (New York: Columbia University Press, 1952), pp. 29–48.

39. Daniel, pp. 102–3, 107–9; McGlothlin, pp. 160–61; Cook, p. 167.

40. Cook, pp. 174–75; McGlothlin, pp. 172–73; Daniel, pp. 109–10.

41. Watson, *Autobiography*, pp. 271–72; Tucker to Birnbaum, February 21, 1961, Furman University; Furman University Transcripts, 1894–1899, Furman University Archives; McGlothlin, p. 179; *Furman University Catalogue, 1896–1897*, p. 27.

42. Moore's introductory course in psychology included discussion of the "facts and modes of consciousness," including "*immediate knowledge*: cognition . . . *feeling* . . . sensation, emotion . . . *Desire* . . . regulation *volition* . . ." Davis's *Elements of Psychology* was used as the text. Baldwin's *Handbook of Psychology*, Sully's *Outlines*, and Ladd's *Psychology, Descriptive and Explanatory* were used as texts in Moore's senior couse; and Wundt, Scripture's *Thinking, Feeling and Doing*, and Ladd's *Physiological Psychology* were cited as references. The course description included a discussion of "*Intellect*: the apperceptive and rational functions . . . *Feeling* . . . the nervous system and consciousness . . . *Will*: the motor consciousness, motor aspects of ideal feeling." (See: *Furman University Catalogue, 1895–1896*, pp. 23–24; and *1896–1897*, p. 27.) Watson's transcripts are located in the Furman University Archives.

43. Watson, *Autobiography*, pp. 271–72.

44. Ibid.

45. Ibid.; "The Hero of the Twentieth Century," *The Furman Echo* XII (December 1899): 16–17; "College Training," *The Furman Echo* XII (December 1899): 48.

46. O. C. Croxton, "Power of Education," *The Furman Echo* VII (February 1895): 147–49.

47. John B. Watson, *Behaviorism* (New York: W. W. Norton, 1970), p. 280.

48. See: Bledstein, pp. 106–7.

49. Brewer, pp. 4–5; McGlothlin, p. 235; *Catalogue of the Batesburg Graded School* (Columbia, S.C.: R.L. Bryan Company, 1904), p. 19; Watson to Yerkes, September 15, 1913, Yerkes Papers.

50. Watson, *Autobiography*, p. 273; Watson to Harper, July 20, 1900, Presidents' Papers. For a discussion of Southern Baptist attitudes toward the University of Chicago in the 1890s, see: James Clyde Harper, "A Study of Alabama Baptist Higher Education and Fundamentalism, 1890– 1930"(Ph.D. diss., University of Alabama, 1977), p. 56. William Rainey Harper's role in making the University of Chicago the home of the new professionalism is discussed in: Mark Beach, "Professional Versus Professorial Control of Higher Education," *The Emerging University and Industrial America*, ed. Hugh Hawkins (Lexington, Mass.: D.C. Heath, 1970), pp. 102–3.

51. Watson to Harper, July 20, 1900, and A. P. Montague to William Rainey Harper, July 20, 1900, Presidents' Papers.

52. See: David Bakan, "Behaviorism and American Urbanization," *Journal of the History of the Behavioral Sciences* II (January 1966): 5–28. Bakan deals with behaviorism as an aspect of urbanization, but he places Watson's confrontation with urban life and values after his arrival in Chicago. Evidence seems to suggest that "urbanization" was a complex and subtle process that was not confined to metropolitan centers. What brought Watson to Chicago in the first place was his aspiration for professional status and mobility, an attitude firmly rooted in the experience of growing up in a rural community that itself was undergoing transformation.

53. See: Bledstein, pp. 105–20.

CHAPTER TWO

1. William James, *Psychology, Briefer Course* (New York: Fawcett Publications, 1963), p. 407. (Original work published 1890).

2. Robert H. Wiebe, *The Search for Order: 1877–1920* (New York: Hill and Wang, 1967), pp. 111–32. See also: Burton Bledstein, *The Culture of Professionalism: The Middle Class and the Development of Higher Education in America* (New York: W. W. Norton, 1976), pp. 105–20, 159–202; T. J. Jackson Lears,

No Place of Grace: Antimodernism and the Transformation of American Culture 1880–1920 (New York: Pantheon Books, 1981), p. 9.

3. Joesph Jastrow, "American Psychology in the '80's and '90's," *Psychological Review* L (January 1943): 65–67; Samuel W. Fernberger, "The American Psychological Association; A Historical Summary, 1892–1930," *Psychological Bulletin* XXIX (January 1932): 1–89.

4. Dorothy Ross, *G. Stanley Hall: The Psychologist as Prophet* (Chicago: University of Chicago Press, 1972), pp. 3–30, 62–63; [William James], "The Teaching of Philosophy in Our Colleges," *The Nation* XXIII (September 1876): 178–79.

5. Ross, pp. 64–98. Those who studied with Wundt during the 1880s were J. M. Cattell, E. B. Titchener, Frank Angell, E. W. Scripture, C. E. Seashore, H. K. Wolfe, E. A. Pace, C. H. Judd, J. M. Baldwin, and Hugo Münsterberg. See: Randall Collins and Joseph Ben-David, "Social Factors in the Origins of a New Science: The Case of Psychology," *American Sociological Review* XXXIV (August 1966): 451–65. See also: Michael M. Sokal, ed., *An American Student in Europe Eighteen Eighty to Eighteen Eighty-Eight: James McKeen Cattell's Journal and Letters from Germany and England* (Cambridge, Mass.: Massachusetts Institute of Technology Press, 1980).

6. E. G. Boring, *A History of Experimental Psychology* (New York: Appleton-Century-Crofts, Inc., 1950), pp. 316–47; Gardner Murphy, *Historical Introduction to Modern Psychology* (New York: Harcourt, Brace and World, Inc., 1949), pp. 137–73; Duane P. Schultz, *A History of Modern Psychology* (New York: Academic Press, 1969), pp. 43–55.

7. John M. O'Donnell, "The Origins of Behaviorism: American Psychology, 1870–1920" (Ph.D. diss., University of Pennsylvania, 1979), pp. 139–251. Published in 1985 (New York: New York Unversity Press).

8. G. Stanley Hall, "The New Psychology [Part I]," *The Andover Review* III (February 1885): p. 134; G. Stanley Hall, "The New Psychology [Part II]," *The Andover Review* III (March 1885): 239.

9. Ross, pp. 105–7.

10. Ibid., pp. 114–15; Bledstein, p. 68; Wiebe, pp. 118–19; Lears, p. 54. See also: Michael B. Katz, *The Irony of Early School Reform: Educational Innovation in Mid-Nineteenth Century Massachusetts* (Cambridge, Mass.: Harvard University Press, 1968), pp. 115–62.

11. Ross, pp. 126, 136; William James, who began his career at Harvard in physiology, became professor of philosophy in 1885 and changed his title to professor of psychology only in 1889. See: Collins and Ben-David, p. 457.

12. Philip J. Pauly, "Psychology at Hopkins: Its Rise and Fall and Rise and Fall and . . . ," *Johns Hopkins Magazine* XXX (December 1979): 36; Hall, "New Psychology [Part I]," p. 134; Hall, "New Psychology [Part II]," p. 247; Ross, p. 138.

13. Ross, p. 154. Although William James had set up a small demonstration laboratory at Harvard in 1875, Hall's was the first laboratory designed to carry out experimental investigation on a permanent basis. See: Collins and Ben-David, p. 457; Schultz, p. 112; Boring, p. 520.

14. Jastrow, p. 66; Ross, pp. 170–74; Hall, "Editorial Note," *American Journal of Psychology* I (November 1887): 3–4.

15. Ross, pp. 177, 65.

16. Ross, pp. 179–82; Roswell P. Angier, "Another Student's Impression of James at the Turn of the Century," *Psychological Review* L (January 1943): 132–33; William James to Carl Stumpf, February 6, 1887, *The Letters of William James*, ed. Henry James (Boston: Little, Brown and Company, 1926): p. 263.

17. Robert M. Yerkes, "Early Days of Comparative Psychology," *Psychological Review* L (January 1943): 75; Ross, pp. 179–82; James to Hugo Münsterberg, February 21, 1892, *The Thought and Character of William James*, Vol. II, ed. R. B. Perry (Boston: Little, Brown and Company, 1935), p. 139.

18. Samuel W. Fernberger, "The American Psychological Association, 1892–1942," *Psychological Review* L (January 1943): 35; Ross, pp. 180–85; Jastrow, p. 65; Thomas M. Camfield, "The Professionalization of American Psychology, 1870–1917," *Journal of the History of the Behavioral Sciences* IX (January 1973): pp. 67–68; Wayne Dennis and E. G. Boring, "The Founding of the APA," *The American Psychologist* VII (March 1952): 95–97.

19. Murphy, pp. 214–16; Boring, pp. 410–20; Schultz, pp. 63–83.

20. Boring, p. 506.

21. Hall, review of *The Principles of Psychology*, by William James, in *American Journal of Psychology* III (February 1891): 585.

22. William James to Theodore Flourney, December 7, 1896, in James, *Letters*, p. 54; James to James Sully, July 8, 1890, in Perry, p. 113; James to G. Croom Robertson, in Perry, p. 115; James to Mary W. Calkins, September 19, 1909, in Perry, p. 123; Perry, p. 120.

23. James to Flourney, in James, *Letters*, p. 54; James to Münsterberg, June 28, 1906, in Perry, p. 471; Lears, p. 39.

24. E. W. Scripture, *The New Psychology* (New York: Charles Scribner's Sons, 1887), pp. 472–73.

25. Camfield, p. 70. For a description of early psychological laboratories and a discussion of the nature and character of contemporary experimental work, see: Frank M. Albrecht, Jr., "The New Psychology in America: 1880–1895" (Ph.D. diss., Johns Hopkins University, 1960), pp. 128–54.

26. James McKeen Cattell, "The Advance of Psychology," *Science* VIII (October 21, 1898): 533; and "The Progress of Psychology," *The Popular Science Monthly* XLIII (October 1898): pp. 784–85.

27. James McKeen Cattell, "Address of the President before the American Psychological Association, 1895," *Pychological Review* III (March 1896): 144; George Trumbull Ladd, "President's Address before the New York Meeting of the American Psychological Association," *Psychological Review* I (January 1894): 18–21; John Dewey, "Psychology and Social Practice," *Science* XI (March 2, 1900): 332.

28. Cattell, "Address," p. 144.

29. A. Tolman Smith, W. T. Harris, and Hugo Münsterberg, "The Psychological Revival," *Report of the Commissioner of Education for the Year 1893–94*, Vol. I (Washington, D.C.: U.S. Government Printing Office, 1896), pp. 425–67; Ross, pp. 279–308.

30. William James, "Talks to Teachers on Psychology," *Atlantic Monthly* LXXXIII (February 1898): 155–62.

31. Josiah Royce, "The New Psychology and the Consulting Psychologist," *The Forum* XXVI (September 1898): 80–96.

32. Dewey, pp. 321–33.

33. Questions raised by Noah Porter in 1877 in an essay called "Is Psychology a Science?" were still hotly debated at the end of the century. See: Noah Porter, *The Elements of Intellectual Science* (New York: Scribner, Armstrong and Co., 1877), pp. 34–41.

34. George Trumbull Ladd, "Is Psychology a Science?" *Psychological Review* I (July 1894): 392–95. See also: Donald S. Napoli, "The Architects of Adjustment: The Practice and Professionalization of American Psychology, 1920–1945" (Ph.D. diss., University of California at Davis, 1975), pp. 16–47.

CHAPTER THREE

1. Quinquennial Statement, July 1, 1896, quoted in: William M. Murphy and D. J. R. Bruckner, eds., *The Idea of the University of Chicago: Selections from the Papers of the First Eight Chief Executives of the University of Chicago from 1891 to 1975* (Chicago: University of Chicago Press, 1976), p. 77.

2. Bernard Duffey, *The Chicago Renaissance in American Letters: A Critical History* (East Lansing: Michigan State College Press, 1954).

3. Christopher Tunnard and Henry Hope Reed, *American Skyline* (New York: New American Library, 1956), pp. 136–53.

4. Thomas W. Goodspeed, *A History of the University of Chicago* (Chicago: University of Chicago Press, 1916).

5. William Rainey Harper, *The Trend in Higher Education* (Chicago: University of Chicago Press, 1905), p. 327.

6. William Rainey Harper to John D. Rockefeller, November 15, 1888,

quoted in: Richard J. Storr, *Harper's University: The Beginnings* (Chicago: University of Chicago Press, 1966), p. 24; see also: pp. 216–18.

7. Harper to Rockefeller, August 9, 1890, in Ibid., p. 47.

8. Thomas W. Goodspeed, *William Rainey Harper* (Chicago: University of Chicago Press, 1928), pp. 148, 126.

9. Ibid., p. 111.

10. Storr, pp. 136, 43.

11. Ibid., p. 137.

12. Harper, "Higher Education in the West," *Trend*, pp. 147–48.

13. ———, "University Training for a Business Career," *Trend*, pp. 270–71.

14. Laurence R. Veysey, *The Emergence of the American University* (Chicago: University of Chicago Press, 1965), pp. 311–17.

15. Storr, pp. 147–49, 155.

16. Goodspeed, *Harper*, p. 147.

17. Thorstein Veblen, *The Higher Learning in America* (New York: B. W. Hubsch, 1918), pp. 220–25, 227–29, 233–38, 250. Although printed in 1918, much of the material was written in 1906 when Veblen was serving an often stormy term on the faculty at the University of Chicago.

18. Harper, "The University and Democracy," *Trend*, pp. 27–28.

19. For an account of the development of sociology at the University of Chicago, see: Steven J. Diner, "Department and Discipline: The Department of Sociology at the University of Chicago, 1892–1920," *Minerva* XIII (Winter 1975): 514–53.

20. Mary O. Furner, *Advocacy and Objectivity: A Crisis in the Professionalization of American Social Science, 1865–1905* (Lexington: University of Kentucky Press, 1975), pp. 168–83.

21. Emil G. Hirsch, "The American University," *American Journal of Sociology* I (September 1895): 113–31. Hirsch was professor of rabbinical literature at Chicago.

22. Darnell Rucker, *The Chicago Pragmatists* (Minneapolis: University of Minnesota Press, 1969), pp. 4–5, 106.

23. John Dewey, "The Reflect Arc Concept in Psychology," *Psychological Review* III (July 1896): 357–70.

24. William James, "The Chicago School," *Psychological Bulletin* I (January 15, 1904): 1–5.

25. James Rowland Angell, "The Relations of Psychology to Philosophy," *The Decennial Publications of the University of Chicago*, Vol. III (Chicago: University of Chicago Press, 1903), pp. 56–59, 63.

26. ———, *Psychology: An Introductory Study of the Structure and Function of Human Consciousness*, 4th ed. (New York: Henry Holt and Company, 1908), p. iii.

27. Ella Flagg Young, "Scientific Method in Education," *Decennial Publications*, p. 149.

28. Rucker, p. 14.

29. John B. Watson, in *A History of Psychology in Autobiography*, Vol. III ed. Carl Murchison (Worcester, Mass.: Clark University Press, 1936), p. 273.

30. Watson, *Autobiography*, pp. 273, 275. See also Watson's suggestive revelations concerning his abandoned medical career in: John B. Watson, "The Psychology of Wish Fulfillment," *The Scientific Monthly* III (November 1916): 482, 485.

31. Watson, *Autobiography*, pp. 273, 274.

32. Ibid., p. 276.

33. Ibid.

34. Philip J. Pauly, *Controlling Life: Jacques Loeb and the Engineering Ideal in Biology* (Oxford, England: Oxford University Press, 1987).

35. ———, "The Loeb–Jennings Debate and the Science of Animal Behavior," *Journal of the History of the Behavioral Sciences* XVII (1981): 505–15.

36. John B. Watson to Jacques Loeb, January 2, 1914, Jacques Loeb Papers, Manuscript Division, Library of Congress, Washington, D. C.; John B. Watson and Will Durant, "Is Man a Machine?" *The Forum* LXXXII (November 1929): 264; Watson, *Autobiography*, p. 273; Pauly, "Loeb–Jennings Debate," p. 512. It is interesting to note that, ironically, Jacques Loeb provided the model for the character of Max Gottlieb, the embattled advocate of "pure" scientific research, in Sinclair Lewis's Pulitzer Prize–winning novel, *Arrowsmith* (1925).

37. John B. Watson, *Animal Education: An Experimental Study on the Psychical Development of the White Rat, Correlated with the Growth of Its Nervous System* (Chicago: University of Chicago Press, 1903), p. 5; see also the review of *Animal Education* in *The Nation* (February 18, 1904): 13. Watson borrowed the $350.00 to have his thesis published from H. H. Donaldson and was two years in repaying the loan.

38. I am indebted to Professor R. A. Boakes for allowing me to see a prepublication manuscript of his recent book, *From Darwin to Behaviorism* (Cambridge University Press, 1984); material on the use of the white rat was taken from Chapter Six.

39. Watson, *Animal Education*, p. 10.

40. Ibid., p. 120.

41. Ibid., pp. 122, 65–66.

42. Watson, *Autobiography*, pp. 273, 274.

43. T. J. Jackson Lears, *No Place of Grace: Antimodernism and the Transformation of American Culture, 1880–1920* (New York: Pantheon Books, 1981), pp. 49–57.

44. Watson, "Psychology of Wish Fulfillment," pp. 484–85.

Notes

45. ———, *Autobiography*, p. 274. Watson is credited with receiving the first Ph. D. in psychology from Chicago, graduating *magna cum laude*. Helen Thompson (Woolley) graduated *summa cum laude* with her major field in psychology before Watson, but at the time the degree was offered only in philosophy. Her dissertation, "Psychological Norms in Men and Women," was a landmark study in sexual roles, which concluded that sex differences were determined more by social than by biological factors. Her career, perhaps, is illustrative of her findings. Although she taught at Mount Holyoke for a short time, she was never afforded the opportunities for recognition in experimental psychology that were available to her male colleagues. See: Margaret W. Rossiter, *Women Scientists in America: Struggles and Strategies to 1940* (Baltimore: Johns Hopkins University Press, 1982), p. 102.

46. John Dewey to James McKeen Cattell, December 5, 1905, James McKeen Cattell Papers, Manuscript Division, Library of Congress, Washington, D.C., cited hereafter as Cattell Papers.

47. H. H. Donaldson to D.C. Gilman, February 10, 1903; Watson, application for research assistant, February 10, 1903, Watson file, Archives of the Carnegie Institution, Washington, D.C., cited hereafter as Carnegie Archives; Watson, *Autobiography*, pp. 274–75; James R. Angell to Dewey, October, 1903, Watson to T. W. Goodspeed, October 13, 1903, Dewey to Harper, October 14, 1903, Presidents' Papers 1889–1925, University of Chicago Archives, cited hereafter as Presidents' Papers.

48. Angell, "Relations," p. 55.

49. Angell to Harper, May 5, 1904, Presidents' Papers.

50. James Rowland Angell, in *A History of Psychology in Autobiography*, Vol. III, ed. Carl Murchison (Worcester, Mass.: Clark University Press, 1936), pp. 12–20.

51. Angell to Harper, July 21, 1904, Presidents' Papers. See also: note 37; Robert M. Yerkes, "Animal Education," *Journal of Comparative Neurology and Psychology* XIV (March 1904): 70–71.

52. Watson to Robert M. Yerkes, August 26, 1904, Robert M. Yerkes Papers, Yale Medical Library, Yale University, New Haven, Conn., cited hereafter as Yerkes Papers.

53. Angell to Harper, September 23, 1904, Presidents' Papers.

54. John B. Watson, "Some Unemphasized Aspects of Comparative Psychology," *Journal of Comparative Neurology and Psychology* XIV (July 1904): 360–63. Watson had a difficult time preparing this paper. See: Watson to Yerkes, August 16, 1904, Yerkes Papers.

55. John B. Watson, "A Suggested Method in Comparative Psychology" (address delivered to the Section in Genetic and Comparative Psychology at the Congress of Arts and Sciences, St. Louis, Missouri, September 21,

1904), typescript in Carnegie Archives. See: Watson to Yerkes, September 4, 1904, Yerkes Papers.

56. John B. Watson to Adolf Meyer, August 13, 1920, Adolf Meyer Papers, John Broadus Watson Correspondence Series I, Unit 3974, Alan Mason Chesney Medical Archives, Johns Hopkins University, Baltimore, Md., cited hereafter as Meyer Papers.

57. Harold Ickes, *The Autobiography of a Curmudgeon* (New York: Renal and Hitchcock, 1943), pp. 7–8.

58. Interview with Mary Watson Hartley, July 12, 1982, Los Angeles, Calif.

59. Watson to Meyer, August 13, 1920, Meyer Papers. See Chapter 7 for an account of the circumstances surrounding Watson's divorce.

60. Harold Ickes to Watson, February 19, 1905, Harold Ickes Papers, Manuscript Division, Library of Congress, Washington, D.C.

61. Ibid.

62. See: Watson to Harper, February 6, 1906, Henry Pratt Judson to Harper, February 7, 1905, Harper to Watson, February 11, 1905, Angell to Harper, February 15, 1905, Harper to Angell, February 18, 1905, Presidents' Papers. See also: Watson to Harper, May 26, 1905, Harper to Watson, June 1, 1905, William Rainey Harper Personal Papers, University of Chicago Archives.

63. See: Watson to Carnegie Institution, January 10, 1905, James Mark Baldwin to Carnegie Institution, January 14, 1904, Donaldson to Carnegie Institution, January 16, 1905, Angell to Carnegie Institution, January 9, 1905; John B. Watson, review of *Psychologie den Niedensten Tiene*, by Franz Lukas (typescript). All contained in Carnegie Archives.

64. Angell to Harper, February 6, 1905, Presidents' Papers; Watson to Yerkes, May 23, 1905, Yerkes Papers.

65. Angell to Harper, June 22, 1905 (see both letters mailed on that date), Harper to Angell, June 24, 1905, Angell to Harper, June 26, 1905, Presidents' Papers.

66. Watson to Yerkes, June 17, 1905, November 2, 1905, Yerkes Papers.

67. See: Watson to Yerkes, August 9, 1906, Yerkes Papers, for a discussion of Watson's work on imitation. Although Watson presented a paper of the results of those experiments in the fall of 1906, the paper was not published until 1908. See: John B. Watson, "Imitation in Monkeys," *Psychological Bulletin* V (June 15, 1908): 169–78. Watson was responsible for the Vol. III, No. 5 (May 15, 1906) edition of the *Psychological Bulletin*. Watson to Yerkes, November 6, 1906, Yerkes papers.

68. Watson to Yerkes, May 23, 1905, Yerkes Papers.

69. Watson to Yerkes, January 9, 1907, Yerkes Papers; Editorial, *The Nation* LXXXIV (January 3, 1907): 2; "To Prosecute Rat Scientist for Cruelty," New York Evening Journal, January 1, 1907, p. 3, c. 3.

70. Angell to Cattell, January 9, 1907, Cattell Papers. James Mark Baldwin, "Professor Watson's Experiments on Rats Defended," *The Nation* LXXXIV (January 24, 1907): 79–80.

71. John B. Watson, "Kinaesthetic and Organic Sensations: Their Role in the Reactions of the White Rat to the Maze," monograph supplement to *Psychological Review* VIII (May 1907).

72. John B. Watson, "Studying the Mind of Animals," *The World Today* XII (April 1907): 421–22.

73. John B. Watson, "The Need for an Experimental Station for the Study of Certain Problems in Animal Behavior," *Psychological Bulletin* III (May 15, 1906): 149–56. For a highly suggestive discussion of the influence of psychology in contributing to notions of biological determinism and cultural relativism, see: George W. Stocking, Jr., "Lamarckianism in American Social Science, 1890–1915," *Race, Culture and Evolution: Essays in the History of Anthropology* (New York: The Free Press, 1968), pp. 234–69. See also: Watson to Yerkes, April 11, 1907, May 29, 1907, Yerkes Papers; Yerkes to C. B. Davenport, January 2, 1907, C. B. Davenport Papers, American Philosophical Society Library, Philadelphia, Pa., for discussions involving plans for acquiring funding for an experimental station.

74. Watson to Yerkes, February 2, 1907, Yerkes Papers.

75. Watson to Yerkes, May 29, 1907, Yerkes Papers; John B. Watson, "A Resume of a Study of the Behavior of Noddy and Sooty Terns Carried on at Bird Key during the Spring of 1907," *Carnegie Institution of Washington Year Book* No. 6, 1907 (Washington, D.C.: Carnegie Institution, 1907), pp. 120–21.

76. Watson to Yerkes, September 26, 1907, Yerkes Papers; Watson refers to Luther Emmett Holt, whose *Care and Feeding of Children* was widely read. I am indebted to Ben Harris for information on this point.

77. Hartley, interview.

78. Graham J. White and John Maze, *Harold Ickes of the New Deal: His Private Life and Public Career* (Cambridge, Mass.: Harvard University Press, 1985); Watson to Meyer, August 13, 1920, Meyer Papers.

79. Watson to E. B. Titchener, December 19, 1908, E. B. Titchener Papers, Cornell University Archives, Ithaca, N.Y., cited hereafter as Titchener Papers.

80. Angell to Baldwin, January 20, 1908, February 12, 1908, Donaldson to Baldwin, February 13, 1908, Watson to Baldwin, March 1, 1908, the Ferdinand Hamburger, Jr., Archives of the Johns Hopkins University, Office of the President, Series #115 (Department of Psychology), cited hereafter as Hamburger Archives.

81. Yerkes to Titchener, August 24, 1908, Titchener Papers; H.S. Jennings to Yerkes, July 20, 1908, May 8, 1908, Yerkes Papers; Jennings to

Baldwin, February 18, 1908, Donaldson to Baldwin, February 13, 1908, Hamburger Archives.

82. Ira Remsen to Baldwin, February 18, 1908, Remsen to Watson, March 3, 1908, Watson to Baldwin, March 1, 1908, Hamburger Archives.

83. Watson to Yerkes, October 2, 1907, Yerkes Papers.

84. Watson to Yerkes, May 27, 1908, Yerkes Papers.

CHAPTER FOUR

1. John B. Watson To E. B. Titchener, June 7, 1909, E. B. Titchener Papers, Department of Manuscripts and University Archives, Cornell University, Ithaca, N.Y., cited hereafter as Titchener Papers.

2. Philip J. Pauly, "Money, Morality and Psychology at Johns Hopkins University: 1881–1942," typescript, Milton S. Eisenhower Library, Johns Hopkins University, 1974, pp. 7–10. A revised version of this paper has been published as: "Psychology at Hopkins: Its Rise and Fall and Rise and Fall and . . . ," *Johns Hopkins Magazine* XXX (December 1979): 36–41.

3. James Mark Baldwin, *Between Two Wars, 1861–1921: Being Memories, Opinions and Letters Received*, Vol. I (Boston: Stratford Company, 1926), p. 120.

4. John B. Watson, in *A History of Psychology in Autobiography*, Vol. III, ed. Carl Murchison (Worcester, Mass.: Clark University Press, 1936), p. 276; John B. Watson to E. B. Titchener, Titchener Papers; Watson to Robert M. Yerkes, September 29, 1908, Robert M. Yerkes Papers, Yale Medical Library, Yale University, New Haven, Conn., cited hereafter as Yerkes Papers.

5. Pauly, pp. 15, 13; Thomas M. Camfield, "The Professionalization of American Psychology, 1870–1917," *Journal of the History of the Behavioral Sciences* IX (January 1973): 68–70.

6. Pauly, pp. 11–14.

7. Watson to Yerkes, March 13, 1909, Yerkes Papers.

8. Watson to Yerkes, December 14, 1908, Yerkes Papers; Watson to Titchener, December 19, 1908, Titchener Papers.

9. Watson to Titchener, June 7, 1909, Titchener Papers.

10. Watson to Yerkes, September 1, 1909, Yerkes Papers.

11. Watson to Ira Remsen, September 4, 1909, the Ferdinand Hamburger, Jr., Archives of the Johns Hopkins University, Office of the President, Series #115 (Department of Psychology), cited hereafter as Hamburger Archives.

12. Watson to Yerkes, September 5, 1909, September 18, 1909, Yerkes Papers.

13. Watson to Arthur O. Lovejoy, April 18, 1910, John B. Watson Papers, Special Collections, Milton S. Eisenhower Library, Johns Hopkins University, Baltimore, Md.

14. Watson to Remsen, September 4, 1909, Hamburger Archives.
15. John B. Watson, "The New Science of Animal Behavior," *Harper's* CXV (February 1910): 346–53. See also: "His School for Animals," Baltimore Sun, February 9, 1910, p. 9, c. 2.
16. Watson to Yerkes, November 15, 1909, February 6, 1910, Yerkes Papers.
17. Watson to James McKeen Cattell, March 7, 1910, Yerkes Papers.
18. Cattell to William James, April 30, 1910, Hugo Münsterberg Papers, Boston Public Library, Boston, Mass., cited hereafter as Münsterberg Papers.
19. Titchener to Cattell, January 16, 1910, James McKeen Cattell Papers, Manuscript Division, Library of Congress, Washington, D.C., cited hereafter as Cattell Papers.
20. Edwin G. Boring, "Titchener's Experimentalists," *Journal of the History of the Behavioral Sciences* III (October 1967): 315–25. In 1903 Titchener had written to James McKeen Cattell: "I hate and abhor the mixed meeting, where the immortality of the soul alternates with the fluctuations of attention. I am convinced that in such a meeting, it is the word papers that get a hearing, while the experimental papers are set under bad conditions, and are not understood or properly discussed. We need an organization of laboratories, in the country: and for that end we must get together and talk experiments and experimental programmes and nothing else." See: Titchener to Cattell, December 8, 1903, Cattell Papers.
21. Watson to Cattell, March 7, 1910, Cattell Papers.
22. Yerkes to Cattell, April 2, 1910, Watson to Yerkes, March 17, 1910, Yerkes Papers.
23. Cattell to James, April 30, 1910, Münsterberg Papers; Watson to Yerkes, June 5, 1910, Yerkes Papers.
24. James to Watson, May 10, 1910, Titchener Papers.
25. For the account of the stormy negotiations surrounding the inception and final dissolution of the congress, see correspondence among Watson, Titchener, Yerkes, James, and Cattell from 1910–1912 in the Titchener, Yerkes, and Cattell Papers.
26. Watson to Cattell, February 13, 1912, Cattell Papers.
27. John M. O'Donnell maintains that this group (for example: H. L. Hollingworth, Albert Paul Weiss, W. B. Pillsbury, Max Meyer, George F. Arps, and Carl Seashore) provided a consensus if not a constituency for the eventual acceptance of Watson's behaviorism. This "silent majority" of psychologists did not base their acceptance of behaviorism on its specific scientific ideas but upon Watson's "general conceptualization of psychology's purpose and scope." See: John M. O'Donnell, *The Origins of Behaviorism: American Psychology, 1870–1920* (New York: New York University Press, 1985).

28. Watson to Titchener, December 14, 1908, Titchener Papers.

29. Watson to Yerkes, January 4, 1910, January 14, 1910, January 18, 1910, Yerkes to Watson, January 17, 1910, Yerkes Papers.

30. Watson wrote Remsen that although he had "no wish to leave Hopkins" he thought that it would be "of interest" for him to know that the University of Illinois had offered him a position as head of a separate department of psychology with unlimited research assistants and a minimum of teaching duties at a salary of $3,500. When Remsen granted him an increase to $3,500 three months later, Watson modestly protested to Yerkes that he was "ashamed" to take the raise and took pains to give the impression that it came gratuitously and spontaneously from an admiring and grateful administration. Although he was later to continue to complain of his poverty, he jubilantly wrote to Yerkes that he would at last be able to clear his debts and afford a trip to Europe. The latter was especially important because he felt the lack of a European tour to be "embarrassing." See: Watson to Remsen, January 4, 1910, Hamburger Archives; Watson to Yerkes, March 22, 1910, Yerkes Papers.

31. Watson to Yerkes, July 17, 1910, Yerkes Papers.

32. Alfred G. Mayer to R. S. Woodward, November 22, 1909, December 25, 1909, December 19, 1911, December 20, 1911, Watson file, Archives of the Carnegie Institution, Washington, D.C.

33. For information on Watson's research in the Dry Tortugas, see the reports on research at Bird Key in the *Carnegie Institution of Washington Year Book*, Nos. 6, 8, 9, 11, 12, and 14, as well as Cedric Larson, "John B. Watson's Research Work for the Carnegie Institution of Washington in the Dry Tortugas" (Paper presented at the Ninth Annual Meeting of Cheiron: The International Society for the History of the Behavioral and Social Sciences, June 11, 1977, University of Colorado at Boulder); see also: Watson to Yerkes, June 5, 1910, June 17, 1912, June 8, 1913, Yerkes Papers; Watson to Remsen, May 16, 1910, Hamburger Archives. It is interesting to note that on an early trip to the island, Watson wrote to Yerkes that four monkeys brought along for companionship and allowed the freedom of the island exhibited a spontaneity that he had never observed in the laboratory. It was the artificiality of Watson's laboratory experiments that some later critics of behaviorism pointed out. See: Watson to Yerkes, May 29, 1907, Yerkes Papers.

34. John B. Watson, "Instinctive Activity in Animals," *Harper's* CXXIV (February 1912): 376–82.

35. Watson to Hugo Münsterberg, September 26, 1911, Münsterberg Papers.

36. Watson, "Instinctive Activity," pp. 376–82.

37. Watson to Yerkes, May 21, 1912, Yerkes Papers; Watson to

Remsen, May 23, 1912, Watson to Administrative Committee, Johns Hopkins University, March 25, 1913, Hamburger Archives.

38. Watson to Meyer, November 14, 1912(?), Adolf Meyer Papers, John Broadus Watson Correspondence Series I, Unit 3974, Alan Mason Chesney Medical Archives, Johns Hopkins University, Baltimore, Md.

39. Christian A. Ruckmich, "The History and Status of Psychology in the United States," *American Journal of Psychology* XXIII (October 1912): 517–31.

40. Watson to Yerkes, October 29, 1909, Yerkes Papers. Watson's textbook, *Behavior: An Introduction to Comparative Psychology*, was not published until 1914 and was the first full-length statement of his behavioristic position, but, as his correspondence suggests, his ideas along this line were already well developed at a much earlier date.

41. Yerkes to Titchener, November 12, 1909, Titchener Papers.

42. Watson to Yerkes, November 30, 1908, Yerkes Papers.

43. Watson to Yerkes, February 6, 1910, Yerkes Papers.

44. Watson to Yerkes, April 7, 1913, Yerkes Papers.

CHAPTER FIVE

1. John B. Watson to Robert M. Yerkes, February 7, 1916, Robert M. Yerkes Papers, Yale Medical Library, Yale University, New Haven, Conn., cited hereafter as Yerkes Papers.

2. See the discussion of the Armory Show in: Henry May, *The End of American Innocence* (Chicago: Quadrangle Books, 1964), pp. 244–47.

3. John B. Watson to Robert M. Yerkes, March 12, 1913, Yerkes Papers.

4. John B. Watson, "Psychology as the Behaviorist Views It," *Psychological Review* XX (March 1913): 158–77.

5. Ibid., p. 158.

6. Ibid., pp. 159–60.

7. Ibid., p. 161.

8. Ibid., p. 163.

9. Ibid., pp. 165–67.

10. Ibid., p. 168.

11. Ibid., pp. 168–69.

12. James McKeen Cattell, "The Conceptions and Methods of Psychology," *Congress of Arts and Science, Universal Exposition, St. Louis, 1904*, Vol. V, ed. Howard J. Rogers (Boston: Houghton Mifflin and Company, 1906), pp. 593–604.

13. James R. Angell, "Behavior as a Category of Psychology," *Psychological Review* XX (July 1913): 255.

14. The argument has been made that, despite its sensational impact, few psychologists were willing to accept Watsonian behaviorism whole-heartedly. See: Franz Samelson, "The Struggle for Scientific Authority: The Reception of Watson's Behaviorism, 1913–1920," *Journal of the History of the Behavioral Sciences* XVII (July 1981): 399–425. Many contemporaries, however, were still able to call themselves "behaviorists" without accepting all of Watson's claims. Bertrand Russell, for example, could accept behaviorism as a method without accepting it as a theory (see Chapter 9). But Watson's call for psychology to be defined in terms of its usefulness struck a responsive chord among many psychologists. John M. O'Donnell convincingly traces the development of this consensus and supports the suggestion that the persuasiveness of behaviorism depended not only on its acceptance as a method or as a psychological theory but also on Watson's articulation of psychology's purpose. See: John M. O'Donnell, *The Origins of Behaviorism: American Psychology, 1870–1920* (New York: New York University Press, 1985), pp. 179–243.

15. John Dewey, "Psychological Doctrine and Philosophical Teach-ing," *Journal of Philosophy, Psychology and Scientific Methods* XI (September 10, 1914): 508–11.

16. Mary Whiton Calkins, "Psychology and the Behaviorist," *Psycholog-ical Bulletin* X (June 15, 1913): 288–91.

17. E. B. Titchener, "On 'Psychology as the Behaviorist Views It'," *Proceedings of the American Philosophical Society* LIII (January–May, 1914), pp. 1–17. James Rowland Angell agreed with much of Titchener's criticism of his former student. He joined with Titchener in a patronizing attitude toward Watson. Assuming an avuncular manner that seemed to look kindly but exasperatedly on a bright but willful youngster, his is perhaps indicative of the attitude of an older generation of psychologists toward Watson and his behavioristic challenge. "You," he wrote to Titchener, "have let Watson down more amiably than most persons will think he deserved. My own disposition would have been to poke a little fun on the historical side of the case. Indeed, I think if Watson had ever had my historical courses, which were developed after he graduated, he would hardly have fallen into some of the pits which have entrapped him. My own position has been rendered rather difficult by virtue of my personal attachment. Of course I am wholly impatient of his position on this issue which seems to me scientifically unsound and philosophically essentially illiterate. . . . For much of his actual work I have very high regard, as I have for him personally. I shall therefore be glad to see him properly spanked, even tho I cannot publicly join in." See: Angell to Titchener, January 25, 1915, E. B. Titchener Papers, Cornell University Archives, Ithaca, N.Y.

18. Watson to Yerkes, October 24, 1916, Yerkes Papers.

19. Brian Mackenzie, "Darwinism and Positivism as Methodological Influences on the Development of Psychology," *Journal of the History of the Behavioral Sciences* XII (October 1976): 330–37.

20. Brian Mackenzie, "Behaviorism and Positivism," *Journal of the History of the Behavioral Sciences* VIII (April 1972): 222–31.

21. John C. Burnham, "On the Origins of Behaviorism," *Journal of the History of the Behavioral Sciences* IV (April 1968): 150.

22. John B. Watson, *Behavior: An Introduction to Comparative Psychology* (New York: Henry Holt and Company, 1914).

23. Walter Lippmann, *Drift and Mastery: An Attempt to Diagnose the Current Unrest* (Englewood Cliffs, N.J.: Prentice Hall, 1961).

24. Ibid., pp. 145–57, 36–44, 52–69. See also: David A. Hollinger, "Science and Anarchy: Walter Lippmann's Drift and Mastery," *American Quarterly* XXIX (Winter 1977): 463–75.

25. For a discussion of the relationship between the ideology of behaviorism and that of progressivism, see: John C. Burnham, "Psychiatry, Psychology and the Progressive Movement," *American Quarterly* XII (Winter 1960): 457–65; Lucille T. Birnbaum, "Behaviorism: John Broadus Watson and American Social Thought, 1913–1933" (Ph.D. diss., University of California at Berkeley, 1964), pp. 113–40.

26. John Dewey, "The Need for Social Psychology," *Psychological Review* XXIV (July 1917): 274.

27. Ibid., pp. 271–73.

28. Ibid., pp. 273–75.

29. Ibid., pp. 275–76.

30. Ibid., p. 276.

31. A suggestive essay on the relation of behaviorism to urban growth is: David Bakan, "Behaviorism and American Urbanization," *Journal of the History of the Behavioral Sciences* II (January 1966): 5–28.

32. John B. Watson, "Image and Affection in Behavior," *Journal of Philosophy, Psychology and Scientific Methods* X (July 31, 1913): 421–28.

33. Watson to Yerkes, February 25, 1915, Yerkes Papers. Despite his criticism, Watson did express a feeling of indebtedness to Loeb and felt that his idea of the reflex and Loeb's notion of tropism were similar concepts. See: John B. Watson to Jacques Loeb, January 2, 1914, Jacques Loeb Papers, Library of Congress, Washington, D.C.

34. Hamilton Cravens, *The Triumph of Evolution: American Scientists and the Heredity–Environment Controversy 1900–1941* (Philadelphia: University of Pennsylvania Press, 1978), p. 208. See also: O'Donnell; H. C. Warren, in *A History of Psychology in Autobiography*, Vol. I, ed. Carl Murchison (Worcester, Mass.: Clark University Press, 1930), p. 462.

35. Watson to James McKeen Cattell, December 22, 1914, James

McKeen Cattell Papers, Manuscript Division, Library of Congress, Washington, D.C., cited hereafter as Cattell Papers.

36. Watson to Yerkes, October 16, 1915, November 9, 1915, Yerkes Papers. Watson published an account of his work in the Tortugas for *Harper's*. See: John B. Watson, "Recent Experiments With Homing Birds," *Harper's* CXXXI (August 1915): 457–64.

37. Watson to Yerkes, October 27, 1915, Yerkes Papers.

38. Yerkes to Watson, October 30, 1915, Watson to Yerkes, November 9, 1915, Yerkes Papers.

39. John B. Watson, "The Place of the Conditioned Reflex in Psychology," *Psychological Review* XXIII (March 1916): 89.

40. Watson gained much of his information on the Russian experiments indirectly. He complained that experimental reports in Russian periodicals were not available and that German and French translations provided only a short sketch of the work. The only material available to Watson in the preparation of his study was Bechterev's summary of his work on conditioned motor reflexes. Ibid., p. 94. Watson later observed that neither Pavlov nor Bechterev "had much influence in shaping my earlier convictions." He "never thought either of them was an objectivist" and "repeatedly pointed out that [Bechterev] was a parallelist at heart." See: Watson to Ernest R. Hilgard, February 18, 1937, Ernest R. Hilgard Papers, Archives of the History of American Psychology, University of Akron, Akron, Ohio, cited hereafter as Hilgard Papers, AHAP. Karl Lashley recalled that "Watson saw [the conditioned reflex] as a basis for a systematic psychology and was not greatly concerned with the nature of the reaction itself." See: Lashley to Hilgard, May 14, 1935, Hilgard Papers, AHAP.

41. Watson also used "violent stimulations" like "burning the subject with a cigarette, tickling with a feather, etc." In addition he used severe electric shock "strong enough to induce perspiration" in the subject. See: Watson, "Place of the Conditioned Reflex," pp. 100, 97.

42. Ibid., pp. 100–1. Watson had not yet determined whether the method could be used successfully with children. An attempt to experiment upon an eight-year-old boy was disappointing for Watson—not to mention the child! "When first punished," Watson related, "he cried and showed some reluctance toward having the experiment continue." Ibid., p. 102.

43. Ibid., p. 101. Laurel Furumoto describes Watson's development of the idea of the conditioned reflex as it later became the basis for a theory of learned behavior. See: Laurel Furumoto, "The Place of the Conditioned Reflex in J. B. Watson's Psychology," typescript, Department of Psychology, Wellesley College, Wellesley, Mass. Watson later remarked that: "It was only later, when I began to dig into the vague word HABIT that I saw the enormous contribution Pavlov had made, and how easily the conditioned response could be looked

upon as the unit of what we had all been calling HABIT." See: Watson to Hilgard, February 18, 1937, Hilgard Papers, AHAP.

44. Watson to Yerkes, February 7, 1916, Yerkes Papers; Nathan G. Hale, Jr., *Freud and the Americans: The Beginnings of Psychoanalysis in the United States, 1876–1917* (New York: Oxford University Press, 1971), p. 447.

45. Watson to Yerkes, November 12, 1914, Yerkes Papers.

46. Watson to Yerkes, February 7, 1916, Yerkes Papers.

47. Ibid.

48. John B. Watson and J. J. B. Morgan, "Emotional Reactions and Psychological Experimentation," *American Journal of Psychology* XXVIII (April 1917): 164.

49. H. S. Jennings to Yerkes, June 10, 1916, Yerkes Papers.

50. Adolf Meyer to Cattell, December 2, 1916, Cattell Papers.

51. Watson to Yerkes, October 12, 1916, Yerkes Papers. See also: Watson to Yerkes, November 25, 1916, Yerkes Papers.

52. Watson to Yerkes, October 18, 1916, November 29, 1916, Yerkes Papers.

53. John B. Watson, "Behavior and the Concept of Mental Disease," *Journal of Philosophy, Psychology and Scientific Methods* XIII (October 26, 1916): 589–96. See also: John B. Watson, "Does Holt Follow Freud?," *Journal of Philosophy, Psychology and Scientific Methods* XIV (February 15, 1917): 85–92; John B. Watson, "The Psychology of Wish Fulfillment," *The Scientific Monthly* III (November 1916): 479–87.

54. Ruth Leys, "Meyer Confronts Watson: Behaviorism and the Origins of American Psychiatry" (Paper presented at the Fifty-third Annual Meeting of the Eastern Psychological Association, Baltimore, Md., April 16, 1982), p. 5.

55. Ibid., p. 6.

56. Meyer to Watson, May 29, 1916, Adolf Meyer Papers, John Broadus Watson Correspondence Series I, Unit 3974, Alan Mason Chesney Medical Archives, Johns Hopkins University, Baltimore, Md., cited hereafter as Meyer Papers.

57. Meyer to Watson, June 3, 1916, Meyer Papers.

58. Meyer to Watson, May 29, 1916, Meyer Papers.

59. Watson to Meyer, June 1, 1916, Meyer Papers.

60. Meyer to Watson, June 3, 1916, Meyer Papers. See also: Philip J. Pauly, "The Loeb–Jennings Debate and the Science of Animal Behavior," *Journal of the History of the Behavioral Sciences* XVII (1981): 505–15.

61. Watson and Morgan, p. 165.

62. Watson, "Behavior and the Concept of Mental Disease," p. 590.

63. Watson and Morgan, pp. 165–69.

64. Watson to Yerkes, March 31, 1916, April 1, 1916, Yerkes Papers.

65. Watson to Jacob H. Hollander, March 21, 1916, Hollander to Watson,

March 22, 1917, April 18, 1917, John B. Watson Papers, Special Collections, Milton S. Eisenhower Library, Johns Hopkins University, Baltimore, Md. Watson's interest in advertising is usually associated with his work for the J. Walter Thompson Company after his dismissal from Johns Hopkins in 1920. Evidence indicates that his interest in this area developed much earlier, coinciding with his work on the conditioned emotional response. Far from being a resort of necessity, it was a natural outgrowth of Watson's interest in developing a methodology that would lead to practical techniques of behavior control.

66. Hugo Münsterberg, *Psychology and Industrial Efficiency* (Boston: Houghton Mifflin Company, 1913). See also: Matthew Hale, Jr., *Human Science and Social Order: Hugo Münsterberg and the Origins of Applied Psychology* (Philadelphia: Temple University Press, 1980).

67. Walter Van Dyke Bingham to C. E. Seashore, March 7, 1916, March 9, 1916, Walter Van Dyke Bingham Papers, Hunt Library, Carnegie-Mellon University, Pittsburgh, Pa., cited hereafter as Bingham Papers.

68. Bingham to Seashore, March 15, 1916, Bingham Papers.

69. John J. Apatow to Bingham, May 15, 1916, Bingham Papers.

70. Bingham to Apatow, May 20, 1916, Bingham Papers.

71. The Economic Psychology Association," undated mimeograph, Hugo Münsterberg Papers, Boston Public Library, Boston, Mass., cited hereafter as Münsterberg Papers.

72. Hugo Münsterberg to Apatow, May 12, 1916, May 25, 1916, Münsterberg Papers.

73. John B. Watson, "The Fake Element in Vocational Psychology," *Baltimore News* (April 26, 1916), clipping in Bingham Papers.

74. John B. Watson, "Attempted Formulation of the Scope of Behavior Psychology," *Psychological Review* XXIV (September 1917): 329–52. Much of this material was incorporated into the first chapter of a book that Watson intended to devote to human psychology. See: John B. Watson, *Psychology from the Standpoint of a Behaviorist* (Philadelphia: J. B. Lippincott Company, 1919).

75. Watson, "Scope of Behavior Psychology," p. 329.

76. Ibid.

77. Ibid., p. 330.

78. Ibid., p. 336.

79. Ibid., pp. 339–40.

CHAPTER SIX

1. Robert M. Yerkes, "Report of the Psychological Committee of the National Research Council," *Psychological Review* XXVI (March 1919): 83.

2. Randolf Bourne, "The Twilight of Idols," *The Seven Arts* II (October 1917): 688–702.

3. David W. Levy, *Herbert Croly of the New Republic: The Life and Thought of an American Progressive* (Princeton, N.J.: Princeton University Press, 1985).

4. John B. Watson to Frank J. Goodnow, March 24, 1917, the Ferdinand Hamburger, Jr., Archives of the Johns Hopkins University, Office of the President, Series #115 (Department of Psychology).

5. David F. Noble, *America by Design: Science, Technology, and the Rise of Corporate Capitalism* (New York: Alfred A. Knopf, 1977), pp. 148–160. The NRC was founded largely through the efforts of George E. Hale, Robert A. Millikan, A. A. Noyes, E. G. Conklin, Charles B. Davenport, and R. S. Woodward. See: Thomas M. Camfield, "Psychologists At War: The History of American Psychology and the First World War" (Ph.D. diss., University of Texas at Austin, 1969), pp. 82–83.

6. George E. Hale, "Origin and Purpose of the National Research Council," memorandum, May 1919, quoted in Noble, p. 154.

7. Walter Dill Scott, "A History of the Committee on Classification of Personnel in the Army," mimeograph, Walter Van Dyke Bingham Papers, Hunt Library, Carnegie–Mellon University, Pittsburgh, Pa., cited hereafter as Bingham Papers.

8. Walter Van Dyke Bingham to James McKeen Cattell, July 30, 1917 (circular letter with enclosures), James McKeen Cattell Papers, Manuscript Division, Library of Congress, Washington, D.C.

9. Watson to Robert M. Yerkes, April 19, 1917, Robert M. Yerkes Papers, Yale Medical Library, Yale University, New Haven, Conn., cited hereafter as Yerkes Papers.

10. James Rowland Angell to Bingham, June 8, 1917, Bingham to Angell, June 13, 1917, Bingham Papers.

11. Angell to Bingham, July 30, 1917, Bingham Papers.

12. A complete treatment of the role of psychologists in World War I is found in: Camfield. A history of the work of Scott's Committee on Classification of Personnel in the Army can be found in: Leonard W. Ferguson, *The Heritage of Industrial Psychology* (Hartford, Conn.: Finlay Press, 1962). Studies of intelligence testing in the army include: Daniel J. Kelves, "Testing the Army's Intelligence: Psychologists and the Military in World War I," *Journal of American History* LV (December 1968): 565–81; Franz Samelson, "World War I Intelligence Testing and the Development of Psychology," *Journal of the History of the Behavioral Sciences* XIII (1977): 274–82; Franz Samelson, "'Putting Psychology on the Map': Ideology and Intelligence Testing," *Psychology in Social Context*, ed. Allan R. Buss (New York: Irvington Publications, 1979), pp. 103–67. Contemporary sources include: U. S. War Department, Committee on Classification of Personnel in the Army, *The Personnel*

System of the United States Army, Vol. I: *History of the Personnel System*, Vol. II: *The Personnel Manual* (Washington, D.C.: U.S. Government Printing Office, 1919); Clarence S. Yoakum and Robert M. Yerkes, *Army Mental Tests* (New York: Henry Holt and Company, 1920).

13. Yerkes to Walter Dill Scott, April 25, 1917, Bingham Papers.

14. Robert M. Yerkes, "Report of the Psychology Committee of the National Research Council," *Psychological Review* XXVI (March 1919): 83. See also: Camfield, pp. 177–95.

15. Samelson, "World War I Intelligence Testing," pp. 276–77.

16. C. B. Davenport to Yerkes, December 18, 1917, C. B. Davenport Papers, American Philosophical Society Library, Philadelphia, Pa., cited hereafter as Davenport Papers.

17. Yerkes to Davenport, December 24, 1917, Davenport Papers.

18. Camfield, pp. 183–217; Kelves, pp. 568–78; Samelson, "World War I Intelligence Testing," p. 279.

19. Camfield, pp. 142–47.

20. Ibid., pp. 150–51.

21. Minutes, Committee on the Classification of Personnel in the Army, August 7, 1917, Bingham Papers. At one point Yerkes considered transferring his program for intelligence testing from the surgeon general's office to that of the Adjutant General but abandoned the effort when he realized the difficulties involved. See: Minutes, Committee on Classification of Personnel in the Army, November 2, 1917, December 5, 1917, Bingham Papers.

22. Scott to the Adjutant General of the Army, August 21, 1917, Walter Dill Scott Papers, Northwestern University Archives, Chicago, Ill.

23. Memorandum, Committee on Classification of Personnel in the Army, September 5, 1917, Bingham Papers.

24. Bingham to Capt. William P. Field, January 4, 1918, Bingham Papers; John B. Watson, in *A History of Psychology in Autobiography*, Vol. III, ed. Carl Murchison (Worcester, Mass.: Clark University Press, 1936), pp. 277–78. For a summary of Watson's work for the Signal Corps and the Military Intelligence Division as well as his work on "shell shock reeducation" see: Yerkes, pp. 97, 138–41, 128–29.

25. Bingham to H. A. Bumstead, September 25, 1918, Bingham to Scott, August 5, 1918, Bingham Papers.

26. Samelson, "Putting Psychology on the Map," pp. 145–47.

27. ———, "World War I Intelligence Testing," p. 276.

28. Camfield, pp. 279–81.

29. James Rowland Angell, in *A History of Psychology in Autobiography*, Vol. III, ed. Carl Murchison (Worcester, Mass.: Clark University Press, 1936), pp. 16–17.

30. Noble, p. 229.

31. Ibid., pp. 229–31; Yerkes, p. 149.

32. Robert C. Clothier to Yerkes, January 20, 1919, Yerkes Papers.

33. Beardsley Ruml to Bingham, April 7, 1919, Bingham Papers.

34. Leonard W. Ferguson, "Industrial Psychology and Labor," *Walter Van Dyke Bingham Memorial Program*, March 23, 1961, ed. Von Haller Gilmer (Pittsburgh: Carnegie Institute of Technology, 1962), pp. 15–21.

35. Ruml to Yerkes, June 21, 1919, Yerkes Papers; Camfield, p. 285.

36. Clothier to Yerkes, June 5, 1923, Yerkes Papers. Beardsley Ruml's administrative career was not atypical. As director of the Laura Spelman Rockefeller Fund he was responsible for establishing the Public Administration Clearing House in Chicago. After serving as dean of the Social Science Department at the University of Chicago, he accepted a position with the R. H. Macy Company in New York, where he became chairman of the board of directors. In addition to serving as an advisor on economic and monetary policy for the Roosevelt administration, he was director of the Federal Reserve Bank of New York. He also served on the boards of (among others) the National Planning Association, the National Bureau of Economic Research, the Market Research Corporation and the National Securities and Research Corporation. See: "Beardsley Ruml," *The National Cyclopaedia of American Biography*, Vol. XLIV, pp. 64–65.

37. Yerkes, pp. 148–49.

38. David Seideman, "Editor with Clout," review of *Herbert Croly of the New Republic*, by David M. Levy, in *The New York Times Book Review*, July 14, 1985, p. 14; Watson to Adolf Meyer, July 5, 1918, Adolf Meyer Papers, Watson Correspondence Series I, Unit 3974, Alan Mason Chesney Medical Archives, Johns Hopkins University, Baltimore, Md.

CHAPTER SEVEN

1. John B. Watson to Rosalie Rayner, April, 1920, Watson v. Watson, No. B680/1920 B-21779 (Circuit Court of Baltimore City, Baltimore, December 24, 1920), letter contained in plaintiff's exhibit.

2. Benedict Crowell, *America's Munitions: 1917–1918* (Washington, D.C.: U.S. Government Printing Office, 1919).

3. Henry May, *The End of American Innocence: A Study of the First Years of Our Own Time, 1912–1917* (New York: Knopf, 1959). See also: David Seideman, "Editor with Clout," review of *Herbert Croly of the New Republic*, by David M. Levy, in *The New York Times Book Review*, July 14, 1985, p. 14.

4. John B. Watson, *Psychology from the Standpoint of a Behaviorist* (Philadelphia: J. B. Lippincott, 1919).

5. Ibid. p. vii.

6. Watson to Frank J. Goodnow, March 30, 1920, Goodnow to Watson, March 31, 1920, the Ferdinand Hamburger, Jr., Archives of the Johns Hopkins University, Office of the President, Series #115 (Department of Psychology), cited hereafter as Hamburger Archives.

7. John B. Watson and K. S. Lashley, "A Consensus of Medical Opinion upon Questions Relating to Sex Education and Venereal Disease Campaigns," *Mental Hygiene* IV (October 1920): 769, 819; Watson to Goodnow, April 29, 1919, Hamburger Archives.

8. For a discussion of "The Socialization of Reproduction and the Collapse of Authority," see: Christopher Lasch, *The Culture of Narcissism: American Life in an Age of Diminishing Expectations* (New York: W. W. Norton and Company, 1978), pp. 154–86.

9. John B. Watson, "Practical and Theoretical Problems in Instinct and Habit," *Suggestions of Modern Science Concerning Education*, H. S. Jennings et al. (New York: Macmillan, 1917) pp. 77–82.

10. Ibid., pp. 73–76.

11. Also participating in the symposium were biologist Herbert Spencer Jennings, sociologist William I. Thomas, and psychiatrist Adolf Meyer.

12. Watson, "Practical and Theoretical Problems," pp. 53–55.

13. Ethel Sturges Dummer to Watson, August 20, 1920, Ethel Sturges Dummer Papers, Schlesinger Library, Radcliffe College, Cambridge, Mass., cited hereafter as Dummer Papers.

14. Watson to Dummer, August 24, 1920, Dummer Papers.

15. Adolf Meyer to Dummer, March 21, 1919, Dummer Papers.

16. Watson to Bertrand Russell, February 21, 1919, Bertrand Russell Papers, Mills Memorial Library, McMaster University, Hamilton, Ontario, cited hereafter as Russell Papers. Enclosed with this letter is a six-page typescript of topics discussed by Watson in one of his staff meetings with Adolf Meyer.

17. Watson to Russell, October 4, 1919, Russell Papers.

18. John B. Watson, "Is Thinking Merely the Action of Language Mechanisms?" *The British Journal of Psychology* XI (October 1920): 87–104.

19. Ibid., pp. 87–89.

20. Ibid., pp. 94, 90.

21. Ibid., pp. 96–103.

22. Lucille T. Birnbaum, "Behaviorism: John Broadus Watson and American Social Thought, 1913–1933" (Ph.D. diss., University of California at Berkeley, 1964), p. 186.

23. John B. Watson, "A Schematic Outline of the Emotions," *Psychological Review* XXVI (May 1919): 192–96.

24. Watson to William I. Thomas, July 29, 1920, Dummer Papers.

25. Watson, "Practical and Theoretical Problems," pp. 55–69.

26. Watson to Thomas, July 29, 1920, Dummer Papers; Watson, *Psychology from the Standpoint of a Behaviorist*, pp. 1–2, 4; T. J. Jackson Lears, *No Place of Grace: Antimodernism and the Transformation of American Culture, 1880–1920* (New York: Pantheon Books, 1981), p. xi.

27. Watson, "Practical and Theoretical Problems," pp. 55–69.

28. Ibid., pp. 77–82.

29. John B. Watson, "The Place of the Conditioned Reflex in Psychology," *Psychological Review* XXIII (March 1916): 89–117.

30. John B. Watson and J. J. B. Morgan, "Emotional Reactions and Psychological Experimentation," *American Journal of Psychology* XXVIII (April 1917): 163–74.

31. John B. Watson and Rosalie Rayner, "Conditioned Emotional Reactions," *Journal of Experimental Psychology* III (February 1920): 1–14.

32. Ibid., pp. 13–14.

33. Ben Harris, "Whatever Happened to Little Albert?" *American Psychologist* XXXIV (February 1979): 158. See also: Franz Samelson, "J. B. Watson's Little Albert, Cyril Burt's Twins, and the Need for a Critical Science," *American Psychologist* XXXV (July 1980): 619–25.

34. Harris, p. 151.

35. Goodnow to Watson, March 18, 1920, Hamburger Archives.

36. See Chapter 3.

37. Watson to James Rowland Angell, August 20, 1910, James Rowland Angell Personal Papers, Yale University, New Haven, Conn. See also: Watson to Meyer, August 13, 1920, Adolf Meyer Papers, John Broadus Watson Correspondence Series I, Unit 3974, Alan Mason Chesney Medical Archives, Johns Hopkins University, Baltimore, Md., cited hereafter as Meyer Papers.

38. See: "William Solomon Rayner," *Baltimore: Its History and Its People* (New York: Lewis Historical Publishing Company, 1912), pp. 879–81; "Rosalie Rayner Watson," Baltimore *Sun*, June 19, 1935; "Albert W. Rayner," Baltimore *Evening Sun*, November 16, 1928.

39. Watson v. Watson, No. B680/1920 B-21779 (Circuit Court of Baltimore City, Baltimore, Md., December 24, 1920), letter contained in plaintiff's exhibit.

40. Watson to Meyer, August 13, 1920, Meyer Papers. See also: Ibid.

41. Ibid.; Watson v. Watson.

42. Meyer to Watson, August 17, 1920, Meyer Papers.

43. Meyer to Goodnow, September 29, 1920, Hamburger Archives.

44. Ibid.

45. Philip J. Pauly, "Money, Morality and Psychology at Johns Hopkins University: 1881–1942," typescript, Milton S. Eisenhower Library, Johns Hopkins University, 1974, p. 19.

46. Meyer to Goodnow, September 29, 1920, Hamburger Archives; Goodnow to Meyer, September 29, 1920, Meyer Papers.

47. Ruth Leys, "Meyer's Dealings with Jones: A Chapter in the History of the American Response to Psychoanalysis," *Journal of the History of the Behavioral Sciences* XVIII (1981): 445–65.

48. William L. O'Neill, *Divorce in the Progressive Era* (New York: New Viewpoints, 1973), pp. 45, 177.

49. Watson to Goodnow, October (?), 1920, Hamburger Archives.

50. Watson to Goodnow, October 4, 1920, Hamburger Archives.

51. Goodnow to Watson, October 7, 1920, Hamburger Archives. There has been a good deal of speculation on the circumstances of Watson's dismissal from Johns Hopkins. H. W. Magoun's "John B. Watson and the Study of Human Sexual Behavior," *The Journal of Sex Research* XVII (November 1981), pp. 368–78, and, especially, the 4th edition of James V. McConnell's introductory psychology text, *Understanding Human Behavior* (New York: Holt, Rinehart & Winston, 1982), p. 284, have accepted as fact the innuendo that Watson was dismissed because of alleged sex research. Their arguments contend that Watson had been engaged in clandestine experiments on female sexual responses with his then assistant, Rosalie Rayner, as his subject. It is further alleged that Watson's wife discovered the records of the experiments, which she used as evidence in her divorce suit.

The evidence that Magoun and McConnell use to support their speculations is based on hearsay (a conversation with one of Watson's advertising colleagues in 1958—the year Watson died and was no longer available to refute the story) and on the existence of a cigar box of instruments (whose functions, Magoun notes, are admittedly open to question) retrieved from the basement of Johns Hopkins University in 1945.

The contention that Mary Ickes Watson found records of sexual experiments that she used as evidence in her divorce suit is simply not true. The complete transcripts of Watson's divorce trial, together with copies of all evidence and testimony presented in the plaintiff's exhibit, contain no instance in which Watson's professional activities were called into question. Furthermore, members of Watson's family completely dismiss the story. Watson's daughter by his first marriage was 15 at the time of the divorce and was well aware of the particulars. His son by his second marriage insists that his parents (who insisted that their children know everything about the "facts of life") never referred to anything vaguely resembling McConnell's story.

It is significant that a close examination of the private correspondence among the university officials and faculty involved in Watson's dismissal does not reveal a single instance in which Watson's professional conduct was criticized. On the contrary, his achievements were highly praised and he was dismissed with great reluctance.

As to Watson's sex research, Magoun notes that Watson, in his 1913 article, "Image and Affection," does mention the "objective registration" of affective responses as a possibility and he speculated on the measurement of the tumescence and detumescence of sex organs as a means of registration. This was written seven years before his dismissal and, though published in a major interdisciplinary journal, provoked no controversy at all; it was certainly no hindrance to his professional career, since he was elected president of the American Psychological Association two years later. Whether Watson actually attempted to measure these responses is open to doubt. The only evidence is a collection of instruments whose function is admittedly open to question and whose attribution to Watson's ownership is purely hearsay. Even if ownership could be proven and Watson's sex research established, one cannot infer that Watson was dismissed because of it. As the confidential correspondence of Watson's colleagues indicates, his professional integrity was never challenged. The explanation for his dismissal lies in the fact that Watson's personal conduct was completely at odds with the expectations of the university, whose officials considered themselves to be custodians of an embattled morality.

52. See Morris Janowitz's introduction to: W. I. Thomas, *On Social Organization and Social Personality* (Chicago: University of Chicago Press, 1966), pp. xiv–xv.

53. "J. B. Watson Sued for Absolute Divorce," Baltimore *Sun*, November 27, 1920, p. 18; "Mrs. Watson Exhibits Letters to 'Rosalie'," Baltimore *Evening Sun*, November 27, 1920, p. 16; "Dr. Watson's Love Note to 'Rosalie' in Court," Baltimore *Sun*, November 28, 1920, p. 26; "Dr. Watson, Divorced, to Pay 4,000 a year," Baltimore *Evening Sun*, December 24, 1920, p. 18; "Mrs. Watson Gains Divorce," Baltimore *Sun*, December 26, 1920, p. 17; "Wife Sues Prof. Watson," *New York Times*, November 27, 1920, p. 4; "Ex-Professor Divorced," *New York Times*, December 25, 1920, p. 16; "Educator's Wife Finds Love Notes," Washington *Post*, November 27, 1920.

54. "Mrs. Watson Reveals Name of 'Rosalie'," Baltimore *Sun*, November 29, 1920, p. 16.

55. W. B. Cannon to Robert M. Yerkes, October 25, 1920, Robert M. Yerkes Papers, Yale Medical Library, Yale University, New Haven, Conn., cited hereafter as Yerkes Papers.

56. W. H. Howell to Yerkes, November 13, 1920, Yerkes Papers.

57. E. L. Thorndike to Yerkes, October 26, 1921, Cannon to Yerkes, November 3, 1920, Arthur O. Lovejoy to Yerkes, November 30, 1920, Yerkes Papers.

58. Dummer to Meyer, November 5, 1920, Dummer Papers.

59. Meyer to Dummer, November 9, 1920, Dummer Papers.

60. Meyer to Watson, August 17, 1920, Meyer Papers. Watson defended himself by an appeal to pragmatism. "I would not rob the world of meaning," he replied to Meyer. "My point was to take our humanity and study 'meaning' and values under varying conditions from the total reaction standpoint. Psychology would not attempt to evaluate them but a non-academic ethics would—an ethics which would lay down rules for practice and guidance." See: Watson to Meyer, August 23, 1920, Meyer Papers.

61. Meyer to Dummer, November 9, 1920, Dummer Papers.

62. F. Scott Fitzgerald, "Echoes of the Jazz Age," *The Fitzgerald Reader*, ed. Arthur Mizener (New York: Scribner's, 1963), pp. 324–25.

63. May, p. 177.

64. Thomas to Dummer, January 1, 1921, Dummer Papers.

65. Thomas to Dummer, January 27, 1921, Dummer Papers.

66. Thomas to Dummer, February 8, 1921, Dummer Papers.

67. Thomas to Dummer, February 11, 1921, Dummer Papers.

68. Thomas to Dummer, May 3, 1920, Dummer Papers.

69. Thomas to Dummer, December 11, 1920, January 1, 1921, Dummer Papers.

70. Mildred V. Bennett to S. I. Franz, November 30, 1920, John B. Watson File, J. Walter Thompson Company Archives, New York, N.Y., cited hereafter as Thompson Archives.

71. E. B. Titchener to Messrs. J. Walter Thompson Company, December 2, 1920, Thompson Archives.

72. Thomas to Dummer, January 1, 1921, Dummer Papers; "Dr. Watson Licensed to Wed Miss Rayner," Baltimore *Sun*, January 2, 1921, p. 26.

CHAPTER EIGHT

1. John B. Watson to Adolf Meyer, August 13, 1920, Adolf Meyer Papers, John Broadus Watson Correspondence Series I, Unit 3974, Alan Mason Chesney Medical Archives, Johns Hopkins University, Baltimore, Md., cited hereafter as Meyer Papers.

2. Stephen J. Fox, *The Mirror Makers: A History of American Advertising and Its Creators* (New York: Vintage/Random House, 1985), p. 79.

3. Ibid., pp. 83–85.

4. Ibid., p. 84.

5. John B. Watson, in *The History of Psychology in Autobiography*, Vol. III, ed. Carl Murchison (Worcester, Mass.: Clark University Press, 1936), p. 280.

6. Watson to Mildred V. Bennett, January 20, 1921, John B. Watson

File, J. Walter Thompson Company Archives, New York, N.Y., cited hereafter as Thompson Archives.

7. Watson to Bertrand Russell, October 11, 1921, Bertrand Russell Papers, Mills Memorial Library, McMaster University, Hamilton, Ontario, Canada; Watson to Walter Van Dyke Bingham, April 16, 1921, Walter Van Dyke Bingham Papers, Hunt Library, Carnegie–Mellon University, Pittsburg, Pa., cited hereafter as Bingham Papers.

8. Watson, *Autobiography*, p. 280.

9. Watson to Meyer, April 9, 1921, Meyer Papers.

10. Stuart Ewen, *Captains of Consciousness: Advertising and the Social Roots of the Consumer Culture* (New York: McGraw-Hill, 1977), pp. 32–33. For a history of psychology in advertising prior to 1920, see: David P. Kuna, "The Psychology of Advertising, 1896–1916" (Ph.D. diss., University of New Hampshire, 1976). See also: Otis Pease, *The Responsibilities of American Advertising: Private Control and Public Influence, 1920–1940* (New Haven, Conn.: Yale University Press, 1958); Fox, p. 94.

11. Association of National Advertisers, Inc., Committee on Research to Bingham, October 24, 1921, Bingham Papers.

12. John B. Watson, "Influencing the Mind of Another" (Speech delivered to the Montreal Advertising Club, September 26, 1935, and reprinted by the J. Walter Thompson Company), John Broadus Watson Papers, Manuscript Division, Library of Congress, Washington, D.C., cited hereafter as Watson Papers.

13. John B. Watson, "The Ideal Executive" (Speech delivered to Macy's graduating class of young executives, April 20, 1922), typescript, Watson Papers.

14. John B. Watson, "Dissecting the Consumer—An Application of Psychology to Advertising," undated typescript, pp. 14, 19, Watson Papers.

15. Watson, "Influencing the Mind of Another."

16. Merle Curti, "The Changing Concept of 'Human Nature' in the Literature of American Advertising," *Business History Review* XLI (Winter 1967): 337–45; David P. Kuna, "The Concept of Suggestion in the Early History of Advertising Psychology," *Journal of the History of the Behavioral Sciences* XII (October 1976): 350–51.

17. Curti, pp. 345–53. In an interview three years before his death, Watson recalled to John C. Burnham that it was not until the 1940s that the advertising industry as a whole became receptive to the suggestions of psychologists (Pease, p. 171n). Otis Pease, however, has set the widespread acceptance of psychology by advertising as beginning with the Depression (Pease, p. 170), while Merle Curti maintains that the estimation of psychology's effectiveness grew among advertisers during the 1920s and was established as the dominant viewpoint by 1930 (Curti, p. 353). The J. Walter

Thompson Company was certainly an industry leader in employing Watson in 1920. Watson's high visibility and success with that firm no doubt was a significant factor in convincing other advertisers to follow Stanley Resor's example. Bernays quoted in: T. J. Jackson Lears, "From Salvation to Self-Realization: Advertising and the Therapeutic Roots of the Consumer Culture, 1880–1930" *The Culture of Consumption: Critical Essays in American History, 1880–1980*, ed. Richard Wrightman Fox and T. J. Jackson Lears (New York: Pantheon/Random House, 1983), p. 20.

18. John B. Watson, "Newspapers and How to Advertise in Them," undated typescript, Watson Papers.

19. "Dr. John B. Watson Favors Testimonial Advertising," *The American Press* (November 1928), p. 14, newspaper clipping, Watson Papers.

20. See Lears, p. xiii. Passage from *Babbitt* quoted in Fox, p. 95.

21. James Rorty, *Our Master's Voice: Advertising* (New York: The John Day Company, 1934), p. 233.

22. "Testimonial Advertising."

23. Pease, pp. 51–56.

24. "Believes Coffee Only Beneficial Stimulant," Baltimore *Sun*, May 3, 1927.

25. John B. Watson, "Advertising by Radio," *J. Walter Thompson Company News Bulletin* No. 98 (May 1923), pp. 11–16.

26. John B. Watson, "What Cigarette Are You Smoking and Why?" *J. Walter Thompson Company News Bulletin* No. 88 (July 1922), pp. 1–15. This was the beginning of the now familiar "blindfold test" in which a client's product is pitted against brand "X." See Watson's obituary in *The Marketing and Social Research Newsletter of the Psychological Corporation* (Summer 1959), pp. 3–4.

27. Rorty, p. 242.

28. "Testimonial Advertising." See also: Howard Gadlin, "Private Lives and Public Order: A Critical View of the History of Intimate Relations in the U.S.," *The Massachusetts Review* XVII (Summer 1976): 324.

29. John B. Watson, "What, To Whom, When, Where, How are We Selling?" (Speech delivered to the J. Walter Thompson Company class in advertising, October 14, 1924), typescript, Thompson Archives.

30. Robert S. Lynd and Helen Merrell Lynd, *Middletown: A Study in Modern American Culture* (New York: Harcourt, Brace and World, 1929), pp. 134–35.

31. John B. Watson and Rosalie Rayner Watson, *Psychological Care of Infant and Child* (New York: W. W. Norton, 1928), pp. 186, 77.

32. Kenneth Macgowan, "The Adventure of the Behaviorist," *The New Yorker* (October 6, 1928), p. 30.

33. Watson to Robert M. Yerkes, May 2, 1923, Robert M. Yerkes Papers, Yale University Medical Library, New Haven, Conn., cited here-

after as Yerkes Papers. See also: Watson to E. G. Boring, September 28, 1936, November 24, 1936, Boring to Watson, September 30, 1936, October 5, 1936, November 27, 1936, E. G. Boring Papers, Harvard University Archives, Cambridge, Mass., cited hereafter as Boring Papers.

34. Yerkes to James McKeen Cattell, March 29, 1921, James McKeen Cattell Papers, Manuscript Division, Library of Congress, Washington, D.C. See: Kerry W. Buckley, "Behaviorism and the Professionalization of American Psychology: A Study of John Broadus Watson, 1878-1958" (Ph.D. diss., University of Massachusetts at Amherst, 1982), pp. 196–226.

35. Henry C. Link, *The New Psychology of Selling and Advertising* (New York: Macmillan, 1932), pp. vii–xiii.

36. John B. Watson, "The Possibilities and Limitations of Psychology in the Office," typescript, pp. 9–12, Thompson Archives; John B. Watson, "Can Psychology Help in the Selection of Personnel?" *J. Walter Thompson News Bulletin* CXXIX (April 1927), pp. 9–10, 13.

37. Watson, "The Ideal Executive."

38. John B. Watson, "Behavioristic Psychology Applied to Selling," *The Red Barrel* XXIII (June 1934), pp. 20–21.

39. Lears, pp. 8–12; F. Scott Fitzgerald, *The Great Gatsby*, quoted in Fox, p. 79.

40. Boring to Watson, January 20, 1925, Boring Papers. See also: E. B. Titchener to Watson, January 17, 1925, E. B. Titchener Papers, Cornell University Archives, Ithaca, N.Y.; Yerkes to Watson, January 23, 1925, Yerkes Papers.

41. William I. Thomas to Ethel Sturges Dummer, June 6, 1920, Ethel Sturges Dummer Papers, Schlesinger Library, Radcliffe College, Cambridge, Mass.

42. John B. Watson, *Behaviorism*, 2nd. ed. (New York: Norton, 1930), p. 40.

43. John B. Watson, "The Myth of the Unconscious," *Harpers* CLV (September 1927), pp. 502–8.

CHAPTER 9

1. John B. Watson, *Behaviorism* (New York: The People's Institute, 1924), p. 248.

2. William I. Thomas to Ethel Sturges Dummer, November 26, 1926, Ethel Sturges Dummer Papers, Schlesinger Library, Radcliffe College, Cambridge, Mass., cited hereafter as Dummer Papers.

3. John B. Watson, "The Analysis of Mind," *The Dial* LXXII (January 1922): 97–202.

4. John B. Watson to Bertrand Russell, January 5, 1922, Bertrand Russell Papers, Mills Memorial Library, McMaster University, Hamilton, Ontario, Canada, cited hereafter as Russell Papers.

5. Ibid. See: Franz Samelson, "Early Behaviorism, Part 3: The Stalemate of the Twenties" (Paper presented at the Twelfth Annual Meeting of Cheiron: The International Society for the History of the Behavioral and Social Sciences, Bowdoin College, Brunswick, Maine, June 1980); John M. O'Donnell, *The Origins of Behaviorism: American Psychology, 1870–1920* (New York: New York University Press, 1985).

6. Ibid.

7. Watson, "Analysis of Mind," pp. 100–1.

8. Bertrand Russell, "An Essay on Behaviorism: A Defense of the Theory That Psychologists Should Observe Impulses, Rather than Speculate upon the Subconscious," *Vanity Fair* XXI (1923), pp. 47, 96, 98.

9. E. B. Titchener to G. Tschelpanow, October 25, 1924, Edward Bradford Titchener Papers, Cornell University Archives, Ithaca, N.Y., cited hereafter as Titchener Papers.

10. Robert M. Yerkes to William I. Thomas, February 2, 1929, Thomas to Yerkes, February 6, 1929, Robert M. Yerkes Papers, Yale Medical Library, Yale University, New Haven, Conn., cited hereafter as Yerkes Papers.

11. Stephen J. Cross, "Antinomies of Science and Society: The Context of Rockefeller Foundation Policy in the 1930s" (Paper presented at the Fourteenth Annual Meeting of Cheiron: The International Society for the History of the Behavioral and Social Sciences, Newport, R.I., June 24, 1982) pp. 6–7.

12. See Chapter 6.

13. Cross, pp. 9–11.

14. Watson to Patty S. Hill, August 1, 1923. Two copies are contained in a letter from J. E. Russell to Beardsley Ruml, August 2, 1923, Rockefeller Archives, Tarrytown, N.Y. Watson also sent a copy of this letter to Bertrand Russell, and it is contained in the Russell Papers.

15. Watson to H. S. Jennings, January 24, 1924, Herbert Spencer Jennings Papers, American Philosophical Library, Philadelphia, Pa., cited hereafter as Jennings Papers. See also: Watson to Yerkes, August 15, 1923, Yerkes Papers.

16. John B. Watson, "Recent Experiments on How We Lose and Change Our Emotional Equipment," Powell Lecture in Psychological Theory at Clark University, January 17, 1925, in *Psychologies of 1925*, ed. Carl Murchison (Worcester, Mass.: Clark University Press, 1927), pp. 63, 65. See also: Mary Cover Jones, "The Elimination of Children's Fears," *Journal of Experimental Psychology* VII (1924): 382–90.

17. Peter M. Rutkoff and William B. Scott, *New School: A History of the New School for Social Research* (New York: The Free Press, 1986); Watson to Yerkes, January 7, 1922, Yerkes Papers.

18. Watson to Jennings, April 3, 1924, Jennings Papers.

19. Alvin Johnson to Dummer, November 24, 1926, Dummer Papers.

20. Joseph Dorfman, *Thorstein Veblen and His America* (New York: Augustus M. Kelley, 1966), p. 450.

21. *New School for Social Research Announcement* (1922–1923), p. 9; (Summer term, 1923), p. 9; (1923–1924), p. 11; (1924–1925), p. 9.

22. Watson to Jennings, March 29, 1924, Jennings Papers.

23. Watson to Jennings, April 3, 1924, Jennings Papers.

24. Hamilton Cravens, *The Triumph of Evolution: American Scientists and the Heredity–Environment Controversy, 1900–1941* (Philadelphia: University of Pennsylvania Press, 1978), pp. 76–77, 201–9.

25. Ibid.

26. Leon J. Kamin, *The Science and Politics of I.Q.* (New York: The Halsted Press, 1974), pp. 15–30.

27. John B. Watson, "Professor McDougall Returns to Religion," typescript contained in Watson to Titchener, March 28, 1923, p. 5, Titchener Papers.

28. Titchener to Raymond Dodge, April 29, 1923, Titchener Papers.

29. Titchener to Watson, April 3, 1923, Titchener Papers.

30. Watson, "McDougall Returns to Religion," p. 5.

31. Robert Boakes, *From Darwin to Behaviourism* (Cambridge, England: Cambridge University Press, 1984), p. 226.

32. Watson to Adolf Meyer, August 13, 1920, Adolf Meyer Papers, John Broadus Watson Correspondence Series I, Unit 3974, Alan Mason Chesney Medical Archives, Johns Hopkins University, Baltimore, Md., cited hereafter as Meyer Papers. See: Chapters 7 and 3.

33. John B. Watson, "Content of a Course in Psychology for Medical Students" (1912?), carbon copy of an original typescript in the Meyer Papers. For Meyer's comments on Watson's views and Watson's reply, see: Meyer to Watson, January 11, 1912, Watson to Meyer, January 16, 1912, Meyer Papers. See also: Chapter 5.

34. John B. Watson, *The Ways of Behaviorism* (New York: Harper and Brothers, 1928), p. 5. For a study of the development and reception of psychoanalysis in the United States, see: Nathan G. Hale, Jr., *Freud and the Americans: The Beginnings of Psychoanalysis in the United States, 1876–1917* (New York: Oxford University Press, 1971).

35. Watson to H. L. Mencken, September 24, 1923, H. L. Mencken Papers, New York Public Library, New York, N.Y.

36. *New School for Social Research Announcement* (1926–1927), pp. 13–14;

see also: Watson to Meyer, October 2, 1926, December 13, 1926, Meyer Papers. See Chapters 4 and 5.

37. Dummer to Watson, March 25, 1927, May 20, 1927, Dummer Papers.

38. Watson to Dummer, April 12, 1927, Dummer Papers. Watson subsequently published his address in *Harper's*. See: John B. Watson, "The Myth of the Unconscious," *Harper's* CLV (September 1927): 502–8. See also: John B. Watson, "The Place of Kinaesthetic, Visceral and Laryngeal Organization on Thinking," *Psychological Review* XXI (1924): 339–48; John B. Watson, "The Unverbalized in Human Behavior," *Psychological Review* XXXI (1924): 273–80.

39. Ralph Barton Perry, "Psychological Theory," *The Saturday Review of Literature*. Clipping in John Broadus Watson File, Special Collections, Milton S. Eisenhower Library, Johns Hopkins University, Baltimore, Md., cited hereafter as Watson Papers, Johns Hopkins.

40. Watson, *Behaviorism*, p. 105.

41. Ibid., p. 11.

42. Rutkoff and Scott, p. 23.

43. B. F. Skinner, *Particulars about My Life* (New York: Alfred A. Knopf, 1976), pp. 298–311; B. F. Skinner, *The Shaping of a Behaviorist* (New York: Alfred A. Knopf, 1979), p. 10. Skinner believed that Watson was "a great ice-breaker and . . . an important figure." But he considered Watson's most important contribution to be "his early work in ethology on noddy and sooty terns." B. F. Skinner to the author, February 11, 1976.

44. John B. Watson, "The Behaviorist's Utopia," typescript, John B. Watson Papers, Manuscript Division, Library of Congress, Washington, D. C., cited hereafter as Watson Papers, Library of Congress. This article was published as: "Should a Child Have More than One Mother?" *Liberty* (June 29, 1929): 31–35. See also: John B. Watson, "Men Won't Marry Fifty Years from Now," *Cosmopolitan* (June 1929): 71, 104, 106.

45. Watson to Meyer, December 4, 1919, Meyer Papers.

46. Watson, "Behaviorist's Utopia," p. 1.

47. Ibid., p. 2.

48. Ibid.

49. John B. Watson and Rosalie Rayner Watson, *The Psychological Care of Infant and Child* (New York: W. W. Norton and Company, 1928).

50. Ibid., pp. 81–82.

51. Watson, "Behaviorist's Utopia," p. 3.

52. See: Watson to Yerkes, January 22, 1932, Yerkes Papers.

53. Watson, "Behaviorist's Utopia," p. 8.

54. Ibid., p. 5.

55. Lillian G. Genn, "Business Unfits Women for Matrimony Says Dr.

John B. Watson," *Everyweek Magazine* (c. 1931), clipping in Watson Papers, Library of Congress.

56. John B. Watson, "The Weakness of Women," *The Nation* CXXV (July 6, 1927): 10.

57. Watson and Watson, p. 12.

58. Watson, "Behaviorist's Utopia," pp. 17–18, 12.

59. Ibid., pp. 12, 18.

60. Ibid., pp. 7, 13–14.

61. Watson, *Behaviorism*, p. 146.

62. Ibid., p. 147.

63. Watson, "Behaviorist's Utopia," pp. 10, 18–19.

64. Perry London, *Behavior Control* (New York: Harper & Row, 1969), pp. 229–82.

65. John B. Watson, "Why I Don't Commit Suicide," typescript, p. 13, Watson Papers, Library of Congress.

66. John B. Watson, "A Letter from John B. Watson," *Journal of Psychology Club*, Furman University (May 1950), pp. 2–3.

67. Watson, *Behaviorism*, 2nd ed. (New York: Norton, 1930), pp. 44, 248.

68. Samuel Haber, *Efficiency and Uplift: Scientific Management in the Progressive Era 1890–1920* (Chicago: University of Chicago Press, 1964), pp. ix–xii.

69. Ibid., pp. 127, 142–43. See also: Warren Sloat, "Looking Back at 'Looking Backward': We Have Seen the Future and It Didn't Work," *New York Times Book Review*, January 17, 1988, p. 3.

70. Watson, *Behaviorism*, p. 82.

71. Watson and Watson, pp. 15–16.

72. Ibid., pp. 9–10, 186.

73. Katherine Tait, *My Father, Bertrand Russell* (New York: Harcourt Brace Jovanovich, 1975), p. 63.

74. Bertrand Russell, "Behaviorist Education," New York *Sun*, May 12, 1928, p. 6, c. 1.

75. Paula S. Fass, *The Damned and the Beautiful: American Youth in the 1920s* (New York: Oxford University Press, 1977), pp. 102–7.

76. Floyd Dell, *Love in the Machine Age: A Psychological Study of the Transition from Patriarchal Society* (New York: Farrar and Rinehart, 1930), pp. 132–37.

77. See collections of clippings in the Watson Papers, Library of Congress, and Watson Papers, Johns Hopkins.

78. John B. Watson, "Feed Me on Facts," *The Saturday Review of Literature* (June 16, 1928): 966–67.

79. Philip Young, *Ernest Hemingway: A Reconsideration* (New York: Harcourt, Brace & World, 1966), pp. 183–90.

80. See: T. S. Eliot, "Hamlet," in *Selected Essays, 1917–1932* (New York:

Harcourt, Brace and Co., Inc., 1932); David Frail, *The Early Politics and Poetics of William Carlos Williams* (Ann Arbor, Mich.: UMI Research Press, 1987).

81. Cecelia Tichi, *Shifting Gears: Technology, Literature, Culture in Modernist America* (Chapel Hill: University of North Carolina Press, 1987).

82. Van Wyck Brooks, "The Literary Life," *Civilization in the United States*, ed. Harold E. Stearns (New York: Harcourt, 1922), pp. 179–97.

83. Watson, *Behaviorism*, p. 199, 225.

84. See the John B. Watson correspondence in the V. F. Calverton and the H. L. Mencken collections, New York Public Library, New York, N.Y. I am indebted to Professor Robert J. Presbie for pointing out the existence of this material.

85. John B. Watson, "How to Grow a Personality," radio broadcast, National Broadcasting Company, January 16, 1932, 8:45 P.M. EST, reprinted by the University of Chicago Press, 1932.

86. *New York Times*, August 2, 1925, Sec. 3, p. 14.

87. Carl C. Jensen, "The Ways of Behaviorism," *Atlantic Monthly*, clipping in Watson Papers, Johns Hopkins.

88. Stuart Chase, review of *Behaviorism*, by John B. Watson, in New York *Herald Tribune*, June 21, 1925, p. 5.

89. See: Joseph Wood Krutch, *The Modern Temper* (New York: Harcourt, 1929).

90. Harvey Wickham, *The Misbehaviorists: Pseudo-Science and the Modern Temper* (New York: The Dial Press, 1931), pp. 14–15, 57–58.

91. Louis Berman, *The Religion Called Behaviorism* (New York: Boni and Liveright, 1927).

92. Mortimer Adler, "John B. Watson as a Scientist Becomes the Billy Sunday of His Own Evangelical Religion," New York *Evening Post*, June 16, 1928.

93. Horace Kallen, "Behaviorism," *Encyclopedia of the Social Sciences*, 1st ed., pp. 495–98.

94. Willard Harrell and Ross Harrison, "The Rise and Fall of Behaviorism," *Journal of General Psychology* XVIII (1938): 367–421.

95. Ibid., pp. 376, 392, 402. In 1930, Watson made his own evaluation of behaviorism's success: ". . . without behaviorism being overtly accepted, its influence has been profound during the 18 years of existence. To be convinced of this, one needs only to compare the contents of our journals title by title for 15 years before the advent of behaviorism and during the past 15 to 18 years. One needs only to compare the books written before and after. Not only have the subjects studied become behavioristic but the *words of the presentations have become behavioristic*. Today no university can escape the teaching of behaviorism." See: John B. Watson, *Behaviorism*, 2nd ed. (New York: W. W. Norton, 1930), pp. x–xi.

96. Kallen, p. 498.

97. For a discussion of social control linking the growth of psychology and the progressive movement, see: John C. Burnham, "Psychiatry, Psychology and the Progressive Movement," *American Quarterly* XII (Winter 1960): 457–65.

CHAPTER 10

1. Robert M. Yerkes to John B. Watson, January 20, 1932, Robert M. Yerkes Papers, Yale Medical Library, Yale University, New Haven, Conn., cited hereafter as Yerkes Papers.

2. Watson to Yerkes, January 22, 1932, Yerkes Papers.

3. Interviews with: Ruth Lieb, New York, N.Y., January 15, 1976, and Woodbury, Conn., May 8, 1976, Joan D. McClure, New York, N.Y., April 30, 1976, Carlotta T. Watson, South Royalton, Vt., June 5, 1976; Jerry Carson, quoted in a letter from Darrel B. Lucas to the author, October 17, 1987; John B. Watson, "The Weakness of Women," *The Nation* CXXV (July 6, 1927): 10.

4. Interview with James B. Watson, Palos Verdes Estates, Calif., July 11, 1982; John B. Watson, "Feed Me on Facts," *The Saturday Review of Literature* (June 16, 1928): 967.

5. Watson, "Why I Don't Commit Suicide," typescript, John Broadus Watson Papers, Manuscript Division, Library of Congress, Washington, D. C., cited hereafter as Watson Papers.

6. *New York Times Book Review* (June 17, 1928), p. 23.

7. James B. Watson, interview.

8. John B. Watson, *Behaviorism* (New York: The People's Institute, 1924), p. 151.

9. James B. Watson, interview.

10. Ibid.

11. Interview with Mariette Hartley Boyriven, Encino, Calif., July 13, 1982.

12. James B. Watson, interview.

13. Lieb, interview.

14. Karl Lashley to E. G. Boring, October 1, 1954, E. G. Boring Papers, Harvard University Archives, Cambridge, Mass.

15. Lieb, interview; B. F. Skinner, "John B. Watson, Behaviorist," *Science* XIII (January 23, 1959): 198.

16. Ruth Lieb, handwritten note, Watson Papers; Lieb, interview.

17. In a striking review essay, Richard Sennett, some years ago, suggested that the hidden theme in B. F. Skinner's popular writings is a nostalgia

for small-town virtues and values. Although facile comparisons between Watson and Skinner should be avoided, parallels in their roles as popularizers would seem to warrant investigation. See: Richard Sennett, review of *Beyond Freedom and Dignity*, by B. F. Skinner, in *New York Times Book Review*, October 24, 1971.

18. I am indebted to T. J. Jackson Lears's discussion of the development of consumption as a hegemonic "way of seeing" for the twentieth century. See Lears's introduction in: *The Culture of Consumption: Critical Essays in American History, 1880–1980*, eds. Richard Wrightman Fox and T. J. Jackson Lears (New York: Pantheon/Random House, 1983), pp. xiv–xvii. See also: Christopher Lasch, *The Culture of Narcissism: American Life in an Age of Diminishing Expectations* (New York: W. W. Norton, 1978), p. 235.

19. William James, *Psychology, Briefer Course* (New York: Fawcett Publications, 1963), p. 408.

Index